Springer Tracts in Natural Philosophy

Volume 36

Edited by C. Truesdell

Harry F. Tiersten

A Development of the Equations of Electromagnetism in Material Continua

With 40 illustrations

Springer-Verlag
New York Berlin Heidelberg
London Paris Tokyo Hong Kong

Harry F. Tiersten
Department of Mechanical Engineering
Aeronautical Engineering & Mechanics
Rensselaer Polytechnic Institute
Troy, NY 12180-3590
USA

Mathematical Subject Classification Codes: 78A25, 78A30

Library of Congress Cataloging-in-Publication Data

Tiersten, Harry F., 1930–
A development of the equations of electromagnetism in material continua / Harry F. Tiersten.
p. cm.—(Springer tracts in natural philosophy; v. 38)
Includes bibliographical references.
ISBN-13:978-1-4613-9681-9
1. Electromagnetism–Mathematics. 2. Maxwell equations.
I. Title. II. Series.
Qc760.T53 1990
530.1'41—dc20 90-9437

Printed on acid-free paper.

Typeset by Thomson Press, New Delhi, India.

9 8 7 6 5 4 3 2 1

ISBN-13:978-1-4613-9681-9 e-ISBN-13:978-1-4613-9679-6
DOI: 10.1007/978-1-4613-9679-6

To the memory of

my mother

Mabel Solomon Tiersten

and my father

Albert Abel Tiersten

Preface

This tract is based on lecture notes for a course in mechanics that has been offered at Rensselaer Polytechnic Institute on and off for the past twenty years. The course is intended to provide graduate students in mechanics with an understanding of electromagnetism and prepare them for studies on the interaction of the electric and magnetic fields with deformable solid continua. As such, it is imperative that the distinction between particle and continuum descriptions of matter be carefully made and that the distinction between that which is inherently linear and that which is intrinsically nonlinear be clearly delineated. Every possible effort has been made on my part to achieve these ends.

I wish to acknowledge the contributions of a number of students and faculty who attended the lectures over the years and who, by their questions and suggestions, significantly improved some of the sections. This preface would not be complete if I did not point out that my interest in electromagnetism was initiated and my attitude towards the development of the equations was influenced by lectures given by the late Professor R.D. Mindlin at Columbia University in the late nineteen fifties. I would like to thank Professor C. Truesdell for his helpful suggestions, which I feel significantly improved the clarity and readability of the Introduction, and Dr. M.G. Ancona for his comment concerning the clarity of an important point in Sec. 1.1.

I wish to express my appreciation to my wife Helen and my children Linda and Steven for their tolerance and understanding. Finally, I would like to thank Geri Frank for her excellent typing of the manuscript from a rather crude draft.

HFT
Schenectady, New York
October 1989

Contents

CHAPTER 1

Introduction

1.1 General

In recent times treatments of electromagnetism develop the field equations for material continua from Maxwell's equations for free-space by defining the macroscopic field variables as statistical averages over the microscopic particle model of the matter. This averaging procedure is not appropriate because in the microscopic particle description only the field equations for free-space are required because the point particles[1] are external to the field. In fact, along these lines Einstein once remarked[2] that the electron is a foreigner to the field. It is important to recognize that in the microscopic particle description the fields need not be defined in regions occupied by continuous distributions of sources, while in the macroscopic description in material continua the fields must be defined in such regions. In the latter case care must be taken to assure that the defined fields in these regions are consistent with the fundamental laws, i.e., Coulomb's and Ampere's, and the rules of mathematics. In the treatments mentioned above such an examination is not ever made and the field equations in material continua are simply averages written. Indeed, although the resulting field equations are perfectly correct for regions containing macroscopic charge and current densities and, in fact, for regions containing macroscopic polarization and magnetization, the reason for their being correct is not established mathematically.

In the case of the macroscopic charge and current densities nothing of interest occurs, but in the case of polarization and magnetization the respective local fields must be omitted from the macroscopic equations. The importance of the local electric field, which is called the Lorentz local field, is well-known and widely appreciated in microscopic solid state physics, but because the usual explanation of this field is based on physical considerations related to the existence of particles in solids, the mathematical significance is not presented or appreciated. An important consequence of this is that the procedures currently employed in the development of the field equations in material continua imply that macroscopic field theories can be obtained for continuous distributions of higher order quantities, e.g., quadrupole densities. However, a direct consideration of the careful mathematical treatment of dipole densities, which results in the exclusion of local fields from the equations, indicates that the development of a theory of quadrupole and, of course, higher order densities is impossible because the fields cannot be defined inside the region containing a quadrupole density unless larger (infinite) quantities are excluded than the finite ones that are retained. In spite of this numerous macroscopic treatments in the literature refer to and explicitly include quadrupole densities up to a point and some even retain them throughout. An important point to be emphasized is that although the fields external to the region in question can be defined for n-pole densities of any order, they cannot be defined inside the region in question.

To be sure in some of the older texts on electromagnetism, notably Livens[3] and Mason and Weaver,[4] the macroscopic fields inside the source regions of the material continua are carefully defined and the mathematical reason for the existence of the local fields is clearly exhibited. In fact the demonstration employed in this work is influenced strongly by Secs. 61–63 of Liven's book.[3] Nevertheless, I consider it appropriate to exhibit this work for a number of reasons. For one thing, the treatment presented here is, I believe, more systematic than the others. For another, this work seeks to separate that which is inherently linear from that which is intrinsically nonlinear and can be made linear only by assumption, while the others make the usual linear assumptions throughout without being careful to delineate where they are made and their significance. Many important consequences arise as a result of this effort, some of which already appear in the published literature.[5] The linear assumptions are made in the other works because problems cannot be formulated and solved without some such assumptions. In this work the formulation and solution of problems is not of direct interest and only the careful development of the field theory, as distinct from the requisite constitutive theory and attendant (linear or nonlinear) assumptions, is pursued. In addition, the motivation for the development presented here is that it is preliminary to the development of convincing descriptions of the interaction of the electric, magnetic and electromagnetic fields with *deformable* heat conducting material continua from well-defined macroscopic models under a variety of physical circumstances.

1.2 Electrostatics

In the treatment presented here the equations of electrostatics, magnetostatics and electromagnetism in material continua are carefully developed starting with simply stated first principles. First the equations for the scalar electrostatic field are developed starting as usual with Coulomb's law for point charges. The point electric dipole moment is defined as the limit of two point charges in the usual way. Then the expressions for the force and torque on an electric dipole are obtained in the usual manner, after which the expression for the energy of a system of interacting point charges and dipoles is established. This energy, of course, resides in the entire configuration and cannot be localized. Then, by introducing the macroscopic charge and polarization densities, which are required to be bounded and satisfy certain continuity conditions, the equation on the electrostatic charge density is obtained by means of Gauss' Theorem. In doing this it is emphasized that the point charges and dipoles are only models that are introduced for convenience in obtaining the field theory for the material continuum.

From the expression for the force on electric charge density and polarization, the expression for the Maxwell electrostatic stress tensor is obtained. From the expression for the energy of a system of interacting point charges and dipoles the expression for the electrostatic energy for a charged and polarized continuum is obtained. From this latter result an expression for the electrostatic energy density of free-space is established in terms of the locally defined electric field, and it is noted that once this has been done the localized electrostatic energy density of free-space has been defined for the field theory. It is further shown that in the general nonlinear case the electrostatic energy density for material continua is exactly the same as that for free-space, but that in this case there is an additional local material energy density.

The integral forms of the electrostatic field equations and force equations are identified as the equations which yield the derived differential forms when the field variables are differentiable. However, the integral forms are taken to hold even when the field variables are not differentiable, such as across surfaces of discontinuity, which enables the usual boundary (jump) conditions on the field vectors across surfaces of discontinuity to be obtained. In addition, it is noted that the jump discontinuity in the normal component of the Maxwell electrostatic stress tensor across a metal-insulator interface determines a force per unit area acting on the interface. In the course of the development the macroscopic energy density of induced polarization is discussd in some detail along with the macroscopic energy density associated with the local fields that were excluded from the macroscopic field theory. Finally, the relation of the general nonlinear electrostatic equations to the usual linear equations is presented.

1.3 Magnetostatics

The development of the equations for the vector magnetostatic field parallels the development of the equations for the scalar electrostatic field step-for-step, with the exception of the fact that because of the vectorial (or cross-product) nature of the magnetostatic field the development is considerably more cumbersome. The equations of the vector magnetostatic field are developed from the mathematically most convenient form of Ampere's law of force between two elements of circulating current loops. In the usual way it is noted that this force system between elements of two loops does not satisfy the law of action and reaction. However, it is shown that the resultant force between an element of one loop and the entire other loop does satisfy the law of action and reaction, i.e., that the forces are equal and opposite and lie along the same line of action. Since in magnetostatics the currents are stationary, all currents are required to flow in closed loops. The magnetic induction and magnetic vector potential are defined in the usual way.

The expression for the magnetostatic energy is obtained and shown to be related to the resultant force between the two current loops by defining the relative rigid translation without rotation for current loops. This demonstration reveals the kinetic character of the magnetostatic energy.

In analogy with the scalar point charges of electrostatics, vector point currents, i.e., moving point charges, are defined in magnetostatics. Similarly, the defined magnetic induction vector at a current element is taken to be due to all other current elements including those of the same loop. The definition of the magnetic vector potential is similarly generalized and, accordingly, the expression for the energy of a current elment due to all other current elements, including those of the same loop, is taken to be the same as the expression obtained for the energy between loops. These generalizations are obviously essential for developing a field theory of magnetostatics. The point magnetic dipole moment is defined as the limit of a circulating current loop converging to a point in the usual way. From this point on the development parallels that of electrostatics in a very direct way. The expressions for the force and torque on a magnetic dipole are obtained straightforwardly. Also, the expression for the energy of a system of interacting point current elements and point dipoles is established. As in the electrostatic case, this energy resides in the entire configuration and cannot be localized. Then, by introducing the macroscopic current and magnetization densities, which are required to be bounded and satisfy certain continuity conditions, the equation on the magnetostatic steady current density is obtained by means of Gauss' Theorem.

From the expressions for the force on the current density and magnetization, the expression for the magnetostatic Maxwell tensor is obtained. From the expression for the energy of a system of interacting current elements and magnetic dipoles, the expression for the magnetostatic energy for a magnetized continuum carrying current is obtained. From this latter result an expression for the magnetostatic energy density of free-space is established in terms of the locally defined magnetic induction field and it is noted that once this has been done, the local magnetostatic energy density for the field theory has been defined. It is shown that in the general nonlinear case the magnetostatic energy density for material continua is the same as that for free-space, but that in this case there is an additional local material energy density.

The integral forms of the field equations and force equations are identified as the equations which yield the derived differential forms when the field variables are differentiable. The integral forms are taken to hold when the field variables are not differntiable, such as across surfaces of discontinuity, which enables the usual boundary (jump) conditions on the field vectors across surfaces of discontinuity to be obtained. Naturally, the jump discontinuity in the normal component of the magnetostatic Maxwell tensor across a material interface determines a force per unit area acting on the interface. In the course of the development the macroscopic energy density of induced magnetization is discussed in some detail along with the macroscopic energy density associated with the local induction field. Then the relation of the general nonlinear magnetostatic theory to the usual linear equations is presented.

1.4 Electromagnetics

Time dependence is introduced by means of Faraday's law in the form stated by Maxwell. Then, following Maxwell, the treatment generalizes the magnetostatic equation on the density of steady-currents to nonsteady currents by employing the continuity equation on electric charge density and assuming that the electrostatic equation on charge density remains valid in the electrodynamic case. From this procedure Maxwell's electromagnetic field equations are obtained in the general form, i.e., without any linear constitutive assumptions. It is shown that when the equations are specialized to the case of free-space they are, indeed, linear and account for the propagation of electromagnetic waves, which includes light, at a speed determined by the ratio of the magnetostatic-to-electrostatic unit of charge.

Before the expression for the force density exerted by the electric field and magnetic induction vector on charged, conducting, polarized and magnetized continua is obtained, it is noted that in Maxwell's time-dependent generalization of the magnetostatic equation he included a current-type term belonging to the matter, which was not included in the development of magnetostatics. Consequently, the force density resulting from this current-type term is added to the expressions for the force densities from electrostatics and magnetostatics to obtain the expression for the total force density, which I prefer to call[6] the rate of supply of linear momentum from the fields to the matter. From the resulting expression for the force density the expression for the electromagnetic Maxwell tensor in material continua is obtained along with the definition of the electromagnetic momentum. The Maxwell tensor turns out to be just the sum of the electrostatic and magnetostatic Maxwell tensors obtained earlier. The expressions obtained for the Maxwell tensor and the electromagnetic momentum are identical with those obtained much earlier by Livens.[3] However, the derivation is quite different, more direct and straightforward and, I believe, more convincing.

Poynting's theorem is obtained from the general form of Maxwell's equations, i.e., without any linear constitutive assumptions, in the usual manner. However, the resulting form taken by the theorem is quite different, but has been in the literature for some time.[5] The expression for the Poynting energy flux is identical with the usual one, but the energy density is just the sum of the electrostatic and magnetostatic energy densities of free-space and the remaining terms simply appear as rate of working or power terms in the general case. As in the earlier case, the application of the integral forms of Maxwell's equations across material surfaces of discontinuity yield the boundary (or jump) conditions across the discontinuity surfaces.

At this point it is noted that the electromagnetic field equations, even for free-space, are not Galilean invariant and, following Einstein,[7] the treatment takes the speed of light to be constant in vacuum independent of the relative speeds of the source and observer. Two inertial coordinate systems in relative motion are considered in the usual way and the special theory of relativity is developed and the Lorentz transformation is exhibited. Then, following Einstein,[7] the treatment shows that Maxwell's equations are invariant under the space-time Lorentz transformation. From this treatment the transformation laws for all electromagnetic field quantities from one inertial system to another are obtained. The low velocity limits of these transformations are essential for the extension of the description to deformable continua. Once these transformations are available I dispense with special relativity.

The final topic treated in this work is the derivation of the equations for linear circuits from the electromagnetic field equations. This is done in the usual way, i.e., by integrating Faraday's law around each circuit and ignoring Maxwell's displacement current in all portions of the circuit except between capacitor plates, where only the quasistatic time-dependent electrostatic equations are taken to hold, which gives the definition of capacitance. From Ohm's law the resistance (or conductance) arises. The main result of the derivation is the definition of mutual- and self-inductance as geometrical integrals. In the treatment it is shown that although the circuits can be treated as lines for the calculations of mutual inductance, in the calculation of self-inductance the circuit must be treated as a three-dimensional region with bounded three-dimensional current density in order that the self-inductance integrals be finite. The resulting equations are shown to be the Kirchhoff voltage equations. The Kirchhoff current equations are obtained by application of the integral form of the continuity equation across each junction.

I

ELECTROSTATICS

CHAPTER 2

Electric Field Equations in Charged Regions

2.1 The Electric Field

This subject starts with Coulomb's law for two point charges, which says that the force exerted by one charge q^Q at Q on the other q^P at P, as shown in Figure 1, may be written in the form

$$\mathbf{F}^{(PQ)} = \frac{\gamma q^Q q^P \hat{\mathbf{r}}}{r^2}, \tag{2.1.1}$$

where $\hat{\mathbf{r}}$ is a unit vector directed along $\mathbf{r}$ from Q to P and γ is a scalar factor introduced for dimensional purposes. There are a fairly large number of dimensional systems in existence and they are all discussed in just about any text on electromagnetic theory. We will discuss some units in greater detail later on. For now we are interested in electrostatic units (e.s.u) only. In this set of units we take $\gamma = 1$, and then unit charges acting one cm apart produce a repulsive force of 1 dyne. This set of units is due to Gauss. It should be noted that other sets of units turn out to be more practical, different sets for different purposes, but this is not our concern here. Clearly, the force at Q due to the charge at P is equal and opposite to $\mathbf{F}^{PQ}$, or

$$\mathbf{F}^{QP} = -\mathbf{F}^{PQ}. \tag{2.1.2}$$

Clearly, in Gaussian units the dimensions of an electric charge can be written

$$Q = L^{3/2} T^{-1} M^{1/2}. \tag{2.1.3}$$

At this stage the electric field at P due to a charge q^Q at Q is defined as the force that would exist at P if a unit charge, $q^P = 1$, were placed at P. Thus

$$\mathbf{E}^{PQ} \equiv \mathbf{E}(P) = q^Q \frac{\hat{\mathbf{r}}}{r^2} = q \frac{\mathbf{r}}{r^3}, \tag{2.1.4}$$

or in component form

$$E_i = q \frac{x_i}{r^3}, \tag{2.1.5}$$

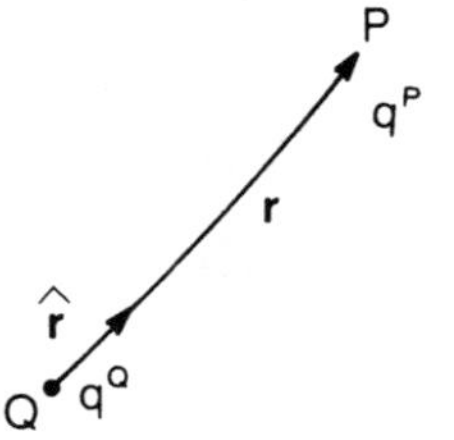

FIGURE 1. Force between two point charges.

but $r^2 = x_k x_k$ and as a result

$$\frac{\partial}{\partial x_i}(r^{-1})_{,i} = -r^{-2} r_{,i} = -r^{-2}\frac{x_i}{r} = -\frac{x_i}{r^3},$$

hence we may write

$$E_i = -\left(\frac{q}{r}\right)_{,i}, \tag{2.1.6}$$

so that we have

$$\mathbf{E} = -\nabla\varphi, \tag{2.1.7}$$

where

$$\varphi = q/r, \tag{2.1.8}$$

and φ is the electric potential. The scalar potential φ can be interpreted as the work done against **E** (or the negative of the work done by **E**) to bring a unit charge from ∞ to the point P in question, as can be seen by considering the diagram in Figure 2 and forming

$$-\int_\infty^P d\mathbf{r}\cdot\mathbf{E} = \int_\infty^P d\mathbf{r}\cdot\nabla\left(\frac{q}{r}\right) = \int_\infty^P dx_i\left(\frac{q}{r}\right)_{,i} = \int_\infty^P d\left(\frac{q}{r}\right) = \frac{q}{r}(P) = \varphi(P). \tag{2.1.9}$$

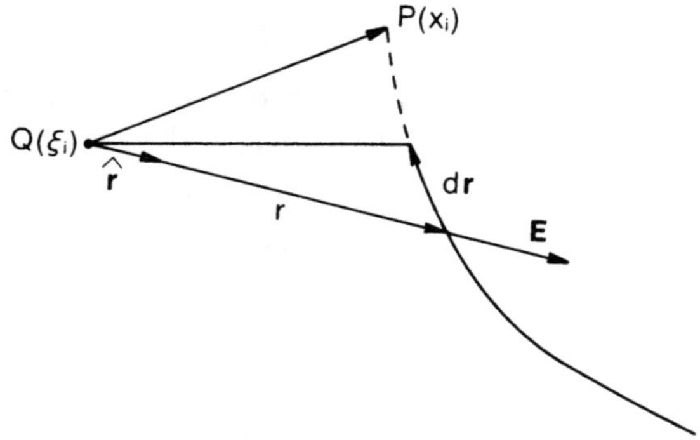

FIGURE 2. Path of unit charge to point P.

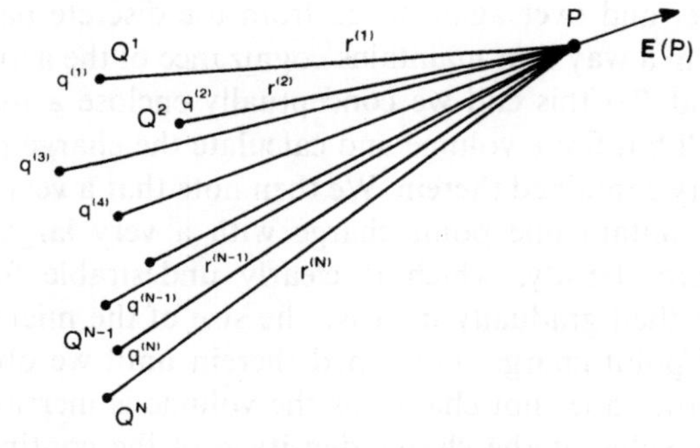

FIGURE 3. System of point charges.

Note that

$$\nabla\cdot\mathbf{E} = -\nabla^2\varphi = -q\nabla^2\left(\frac{1}{r}\right), \tag{2.1.10}$$

and that

$$\nabla^2\left(\frac{1}{r}\right) = 0 \quad \text{except at } r = 0. \tag{2.1.11}$$

Hence, we have

$$\nabla^2\varphi = 0, \quad \nabla\cdot\mathbf{E} = 0 \quad \text{except at } Q. \tag{2.1.12}$$

Now, let us consider a system of point charges Q^n, $n = 1, 2, 3, \ldots N$, all distinct from each other and P as shown in Figure 3, and, by superposition, determine the electric field and electric potential at P due to all the point charges. Thus

$$\mathbf{E}(P) = -\sum_{n=1}^{N} \nabla \frac{q^{(n)}}{r^{(n)}}, \tag{2.1.13}$$

$$\varphi(P) = -\int_{\infty}^{P} d\mathbf{r}\cdot\mathbf{E} = \int_{\infty}^{P} d\mathbf{r}\cdot \sum_{n=1}^{N} \nabla \frac{q^{(n)}}{r^{(n)}}$$

$$= \int_{\infty}^{P} dx_i \sum_{n=1}^{N} \frac{\partial}{\partial x_i}\left(\frac{q^{(n)}}{r^{(n)}}\right) = \sum_{n=1}^{N} \frac{q^{(n)}}{r^{(n)}}, \tag{2.1.14}$$

and clearly we have

$$\nabla^2\varphi = 0, \quad \nabla\cdot\mathbf{E} = 0 \quad \text{except at each } Q^{(n)}. \tag{2.1.15}$$

2.2 Continuous Charge Distributions

We now proceed to replace the discrete point charges by a variable distribution of charge density ρ, which may have discontinuities at points, lines and surfaces. However, in order to do this we introduce an argument that is used over and over again to go from the discrete particle model to the continuous in a way that maintains cognizance of the actual microscopic nature of a solid. To this end we conceptually enclose a number of the Q points in a small but finite volume and calculate the charge per unit volume or charge density contained therein. We then note that a very small but finite volume might contain one point charge with a very large density or no charge with zero density, which is clearly undesirable for defining the continuum. We then gradually increase the size of the microscopic volume and number of point charges contained therein until we obtain a value of charge density that does not change as the volume is increased further. We then define this value as the charge density ρ of the continuum and write $\rho\, dV$ for the charge contained in the differential element of volume of the continuum, which we note approaches zero mathematically and contains a nondenumerable infinity of points, whereas the discrete model contains a finite number of points. In this way we replace $\sum_n q^{(n)}$ at the Q points in the discrete model by the $\int \rho\, dV$ at the Q points in the continuum.[8]

Since we now have two distinct points in the continuum, i.e., the source, or Q, points and the field, or P, points, we must carefully distinguish between them. We do this by representing the coordinates of the P point by x_i and the Q point by ξ_i, as shown in Figure 4, where

$$r_1 = \sqrt{(x_k - \xi_k)(x_k - \xi_k)}, \tag{2.2.1}$$

is the distance between the source point and the field point, and we have $r_1 = r$ when $\xi_k = 0$. Clearly then in accordance with the previous paragraph, when the sums in the previous expressions are replaced by integrals, the $r^{(n)}$ are replaced by r_1 and the $q^{(n)}$ by $\rho\, dV$. Then from (2.1.14) and (2.1.13), respectively, we may write the expressions.

$$\varphi(P) = \int_{V(Q)} \frac{\rho\, dV}{r_1}, \tag{2.2.2}$$

$$\mathbf{E} - \nabla^P \varphi(P) = -\int_{V(Q)} \rho \nabla^P \left(\frac{1}{r_1}\right) dV, \tag{2.2.3}$$

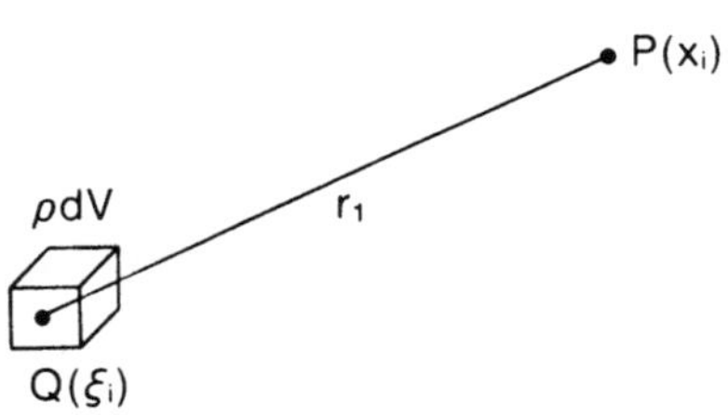

FIGURE 4. Source point and field point in the continuum.

where

$$\nabla^P = \mathbf{e}_k \frac{\partial}{\partial x_k}, \tag{2.2.4}$$

is the gradient at the field point, and we have to carefully distinguish between the gradient at the field point ∇^P and the gradient at the source point ∇^Q, which is given by

$$\nabla^Q = \mathbf{e}_k \frac{\partial}{\partial \xi_k}. \tag{2.2.5}$$

Then it is clear that

$$\nabla^P\left(\frac{1}{r_1}\right) = -\nabla^Q\left(\frac{1}{r_1}\right), \tag{2.2.6}$$

because of the form of r_1. Clearly then, we also have the relations

$$\nabla^P \cdot \nabla^P \varphi \equiv \nabla^2 \varphi = 0 \quad \text{except where } \rho \neq 0, \tag{2.2.7}$$

$$\nabla^P \cdot \mathbf{E} = \nabla \cdot \mathbf{E} = 0 \quad \text{except where } \rho \neq 0. \tag{2.2.8}$$

We have now discussed the electric potential and electric field outside the charged region for a volume distribution of charge density. In a similar way we can discuss the potential and field outside the charged region in the case of a surface distribution of charge density. In this case we replace the discrete point charges by a variable distribution of surface charge σ per unit area and the sums by surface integrals. All of the relations we obtained in the case of the volume distribution of charge density clearly remain valid except for the replacement of the volume integrals by surface integrals. Thus, we have

$$\varphi(P) = \int_{S(Q)} \frac{\sigma\, ds}{r_1}, \tag{2.2.9}$$

and

$$\mathbf{E} = -\nabla^P \varphi(P) = -\int_{S(Q)} \sigma \nabla^P\left(\frac{1}{r_1}\right) dS, \tag{2.2.10}$$

in place of the volume integrals in (2.2.2) and (2.2.3), respectively. It should be noted that in this case we are always outside of the surface charge region because φ and $\mathbf{E}$ satisfy differential equations, which hold in volumetric regions only and have no meaning on a surface. Moreover, it should be clear that in the surface case we have replaced the $q^{(n)}$ by $\sigma\, dS$ and in the volume case by $\rho\, dV$. In addition, it should also be clear that the existence of a nonzero σ on some surface corresponds to a volumetric region in which ρ diverges (becomes infinite). In fact for such a situation we have the pill box shown in Figure 5, which enables us to write

$$\rho\, dV = \rho t\, dS = \sigma\, dS, \tag{2.2.11}$$

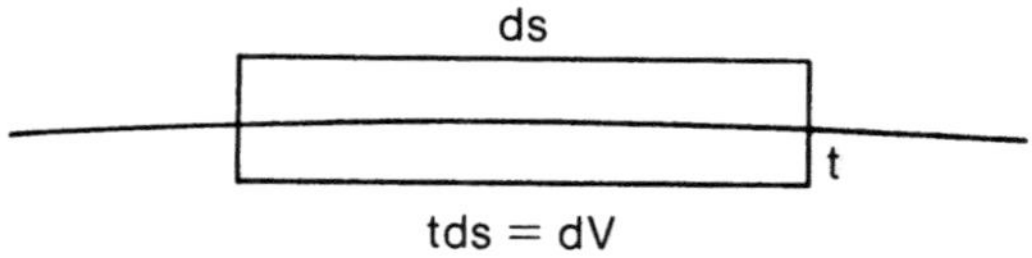

FIGURE 5. Pill-box region.

so that

$$\sigma = \lim_{\substack{t \to 0 \\ \rho \to \infty}} \rho t. \tag{2.2.12}$$

Clearly, if you have surface and volume distributions you simply add the expressions on the right-hand side of Eqs. (2.2.3), (2.2.2) and (2.2.10), (2.2.9), respectively, for **E** and φ.

2.3 Geometric Considerations

We would now like to consider the electric fields and potentials inside the charged region. However, before we can do this we have to have a brief digression in which we discuss two distinct mathematical entities that are very important in the sequel. The first is the concept of the solid angle, which we now consider. To this end we note that the lines from a point 0 through the points of a curve enclosing an area S generate a cone, as shown in Figure 6. The surface area of the unit sphere about 0 intercepted by this cone is called the solid angle Ω of the cone.

Consider the expression

$$\int_S \frac{\mathbf{n} \cdot \mathbf{r}}{r^3} dS = \int_S \frac{\mathbf{n} \cdot \hat{\mathbf{r}}}{r^2} dS = \int_S \frac{\cos\theta\, dS}{r^2}, \tag{2.3.1}$$

which clearly is dimensionless and, hence, has the units of angle in radians; and indeed from its form represents an angle. Thus, we may write

$$d\omega = \frac{\cos\theta\, dS}{r^2}, \tag{2.3.2}$$

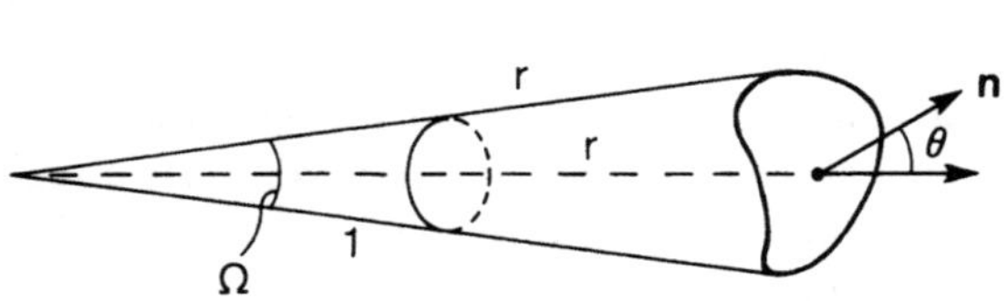

FIGURE 6. Diagram for solid angle.

where $d\omega$ is an infinitesimal increment of solid angle ω, so that

$$\omega = \int_S d\omega. \tag{2.3.3}$$

Since $r \neq 0$, we may apply the divergence theorem to

$$\int_{\mathscr{S}} \frac{\mathbf{n}\cdot\mathbf{r}}{r^3}\, dS, \tag{2.3.4}$$

where $\mathscr{S}$ denotes the surface enclosing the volume bounded by the conical surface, the surface S and the portion of the spherical surface of radius $r = 1$. Thus, we may write

$$\int_S \frac{\mathbf{n}\cdot\mathbf{r}}{r^3}\, dS + \int_{r=1} \frac{\mathbf{n}\cdot\mathbf{r}}{r^3}\, dS = \int_{\mathscr{V}} \nabla\cdot\frac{\mathbf{r}}{r^3}\, dV, \tag{2.3.5}$$

since $\mathbf{n}\cdot\mathbf{r} = 0$ on the conical surface. Moreover, we have

$$\nabla\cdot\frac{\mathbf{r}}{r^3} = \frac{\partial}{\partial x_i}\frac{x_i}{r^3} = \frac{3}{r^3} - x_i 3r^{-4}\frac{x_i}{r} = 0. \tag{2.3.6}$$

In addition on the inner surface $r = 1$, $\mathbf{n} = -\mathbf{r}/r$ and, hence, we have

$$\int_S \frac{\mathbf{n}\cdot\mathbf{r}}{r^3}\, dS = \int d\Omega = \Omega, \tag{2.3.7}$$

and it is clear that $\omega = \Omega$.

If S is a closed surface surrounding 0

$$\Omega = \omega = 4\pi, \tag{2.3.8}$$

because the area of a unit sphere is 4π. However if S is a closed surface and 0 is outside S

$$\Omega = \omega = 0, \tag{2.3.9}$$

because of the diagram shown in Figure 7, which shows that the cone subtends a solid angle ω_0 from the outside portion of the closed surface and $-\omega_0$ from the inside portion, for a total of zero. It is easy to see that the same results hold[9] for closed surfaces such as those shown in Figure 8.

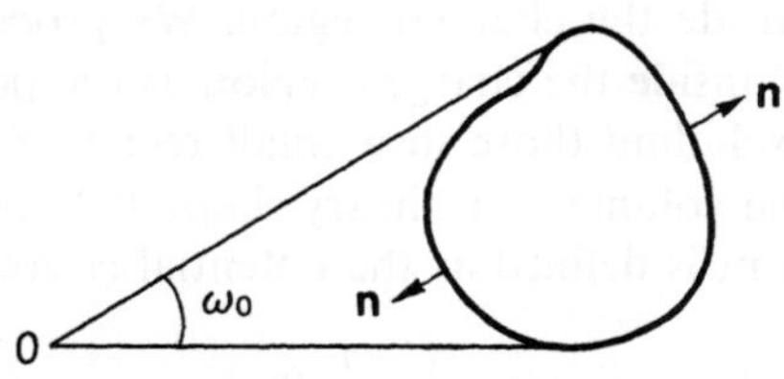

FIGURE 7. Point outside closed surface.

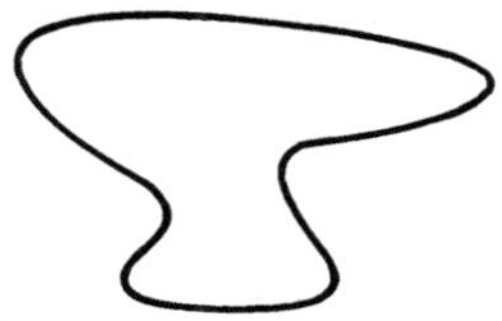

FIGURE 8. Complex closed surface.

This brings us to the consideration of the second entity, which arises from the fact that in the treatment we will be considering integrals which contain integrands $f(r_1)$ that become infinite at $r_1 = 0$. We deal with this situation by splitting the region of integration into two parts, one V' being a region in which f remains bounded and the other V'' being a small region containing all points at which f becomes unbounded. In this latter region it is important for us to know that

$$\int_{V''} \frac{dV}{r^n} \to 0, \tag{2.3.10}$$

for $n < 3$ and $V'' < 4\pi a^3/3$ as $a \to 0$, where a is the radius of a sphere surrounding V''. To show this we simply note that

$$\int_{V''} \frac{dV}{r^n} \leqslant 4\pi \int_0^a \frac{r^2\,dr}{r^n} = 4\pi \left[\frac{r^{3-n}}{3-n}\right]_0^a,$$

since the integral is positive and if $n < 3$, $r^{3-n} = 0$ at $r = 0$ and we have

$$\int_{V''} \frac{dV}{r^n} \leqslant \frac{4\pi}{3-n} a^{3-n}, \tag{2.3.11}$$

which vanishes as $a \to 0$.

2.4 Fields Inside the Charged Region

Consider the expression (2.2.2), i.e.,

$$\varphi = \int_V \frac{\rho}{r_1} dV, \tag{2.4.1}$$

for the potential outside the charged region. At present this expression is not valid for points P inside the charged region. We proceed by defining the potential at points P inside the charged region as the potential φ' at P due to all the charges excluding those in a small region V'' containing P and taking the limit as the volume of arbitrary shape $V'' \to 0$. If φ' approaches a definite limit, that limit is defined as the potential φ. Accordingly, we write

$$\varphi' = \int_{V'} \frac{\rho}{r_1} dV, \tag{2.4.2}$$

and since here $n = 1$, from (2.3.10) it is clear that if ρ is bounded we have

$$\lim_{V''\to 0}\int_{V''}\frac{\rho}{r_1}\,dV\to 0, \tag{2.4.3}$$

and since $V'\to V$ as $V''\to 0$ we may write

$$\varphi = \lim_{V''\to 0}\varphi' = \lim_{V'\to V}\int_{V'}\frac{\rho}{r_1}\,dV = \int_V\frac{\rho}{r_1}\,dV - \lim_{V''\to 0}\int_{V''}\frac{\rho}{r_1}\,dV = \int_V\frac{\rho}{r_1}\,dV, \tag{2.4.4}$$

thus showing that the same expression for φ is valid inside the charged region as outside.

We proceed in exactly the same way to find the expression for **E** inside the charged region. The only difference is that instead of the expression for φ we use the expression (2.2.3) for **E**, i.e.,

$$\mathbf{E}' = -\int_{V'}\rho\nabla^P\left(\frac{1}{r_1}\right)dV, \tag{2.4.5}$$

and since

$$\nabla^P\left(\frac{1}{r_1}\right) = -\frac{1}{r_1^2}\hat{\mathbf{r}}, \tag{2.4.6}$$

we have

$$\left|\nabla^P\left(\frac{1}{r_1}\right)\right| = \frac{1}{r_1^2}, \tag{2.4.7}$$

and again from (2.3.10) it is obvious that if ρ is bounded we have

$$\lim_{V''\to 0}\left|\int_{V''}\rho\nabla^P\left(\frac{1}{r_1}\right)dV\right| < \lim_{V''\to 0}\int_{V''}\frac{|\rho|}{r_1^2}\,dV\to 0, \tag{2.4.8}$$

since here $n = 2$. Thus we may write

$$\begin{aligned}\mathbf{E} &= \lim_{V''\to 0}\mathbf{E}' = -\lim_{V'\to V}\int\rho\nabla^P\left(\frac{1}{r_1}\right)dV\\ &= -\int_V\rho\nabla^P\left(\frac{1}{r_1}\right)dV + \lim_{V''\to 0}\int_{V''}\rho\nabla^P\left(\frac{1}{r_1}\right)dV\\ &= -\nabla^P\int_V\frac{\rho}{r_1}\,dV,\end{aligned} \tag{2.4.9}$$

which shows that the same expression for **E** is valid inside the charged region as outside. In addition, it is perfectly clear from the demonstration, that φ and **E** are continuous and satisfy the equation

$$\mathbf{E} = -\nabla^P\varphi, \tag{2.4.10}$$

inside as well as outside the charge region.

Up to now the charge density has been bounded and integrable. In order for the gradient of the electric field to exist and be continuous at interior points, which we require in the sequel, further restrictions on the charge density are required. Accordingly, we will now show that the first spatial derivatives of electric field and the second spatial derivatives of φ exist and are continuous at interior points if ρ and its first derivatives are continuous. To this end consider the expression

$$\mathbf{E} = -\int_V \rho \nabla^P\left(\frac{1}{r_1}\right) dV, \tag{2.4.11}$$

which, with (2.2.6) may be written

$$\mathbf{E} = \int_V \rho \nabla^Q\left(\frac{1}{r_1}\right) dV. \tag{2.4.12}$$

This last expression may be written

$$\mathbf{E} = \int_V \nabla^Q\left(\frac{\rho}{r_1}\right) dV - \int_V \frac{\nabla^Q \rho}{r_1} dV, \tag{2.4.13}$$

and by a well-known integral theorem this may be written

$$\mathbf{E} = \int_S \mathbf{n}\frac{\rho}{r_1} dS - \int_V \frac{\nabla^Q \rho}{r_1} dV. \tag{2.4.14}$$

Since S is the surface bounding V and the point P is confined to the interior of V, the value of r_1 in the first integral does not vanish, and, hence, that integral has continuous derivatives of all orders. The second integral may be written

$$\mathbf{e}_i \int_V \frac{1}{r_1}\frac{\partial \rho}{\partial \xi_i} dV, \tag{2.4.15}$$

and the coefficient of each base vector is an integral of the same form as

$$\int_V \frac{\rho}{r_1} dV, \tag{2.4.16}$$

which we have just shown has continuous first derivatives. Hence, the expression for $\mathbf{E}$ has continuous first derivatives. Consequently, the divergence theorem

$$\int_S \mathbf{n}\cdot\mathbf{E}\, dS = \int_V \nabla\cdot\mathbf{E}\, dV, \tag{2.4.17}$$

is valid for S being *any* closed surface enclosing a volume V, within which ρ and $\nabla\rho$ are continuous. However, it should be noted that although $\partial E_i/\partial x_j$

exits it may not necessarily be derived from

$$E_i = -\int_V \rho \frac{\partial}{\partial x_i}\left(\frac{1}{r_1}\right) dV,$$

by differentiation under the integral sign. In other words

$$\frac{\partial E_i}{\partial x_j} \neq -\int_V \rho \frac{\partial^2}{\partial x_i \partial x_j}\left(\frac{1}{r_1}\right) dV,$$

necessarily.

2.5 Field Equation on Charge Density

Consider the entire charged region enclosed within S_0 shown in Figure 9 and consider *any* surface S within which the charge and its gradient are continuous. Integrate the normal components of **E** over S to obtain the scalar

$$\int_S \mathbf{n}\cdot\mathbf{E}\, dS, \tag{2.5.1}$$

which with (2.4.4) and (2.4.10) may be written

$$\int_S \mathbf{n}\cdot\mathbf{E}\, dS = -\int_S \mathbf{n}\cdot\left[\left(\int_{V_0} \rho \nabla^P\left(\frac{1}{r_1}\right) dV\right) dS = \int_S \mathbf{n}\cdot\left[\int_{V_0} \rho \frac{\mathbf{r}_1}{r_1^3} dV\right] dS. \tag{2.5.2}$$

Interchanging the order of integration, we obtain

$$\int_S \mathbf{n}\cdot\mathbf{E}\, dS = \int_{V_0} \rho \int_S \mathbf{n}\cdot\frac{\mathbf{r}_1}{r_1^3} dS\, dV. \tag{2.5.3}$$

Integrating the right-hand side of (2.5.3) over the *closed* surface S and employing (2.3.7)–(2.3.9), we obtain

$$\int_S \mathbf{n}\cdot\mathbf{E}\, dS = 4\pi \int_V \rho\, dV, \tag{2.5.4}$$

where V is the volume enclosed by S and V_0, of course, represents the volume

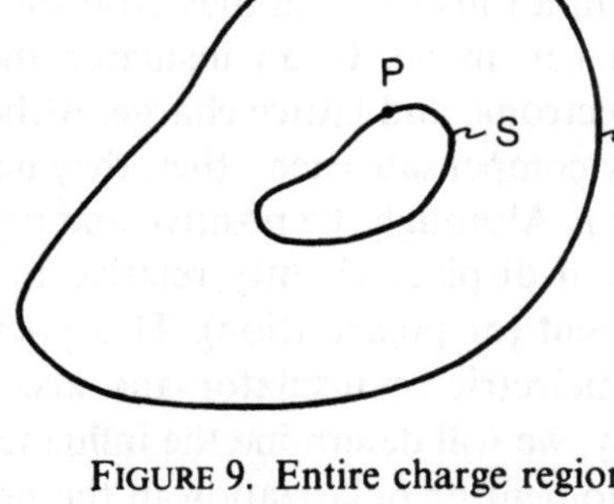

FIGURE 9. Entire charge region.

of the entire charged region. This last expression is a consequence of the fact that, as shown in Sect. 2.3, for a closed surface S, the solid angle for any point outside vanishes and for any point inside is 4π. At this point we note that since the integration coordinates are dummy variables on each side of (2.5.4), we may ignore the P and Q points and replace ξ_i and x_i by, say, y_i in (2.5.4) and in the sequel and, in fact, then let y_i be written as x_i. Since we have restricted ρ and its first derivatives to be continuous within V, the first derivatives of $\mathbf{E}$ are continous, so that the divergence theorem is valid and we have

$$\int_V (\nabla\cdot\mathbf{E} - 4\pi\rho)\, dV = 0, \tag{2.5.5}$$

for *any* volume V including an arbitrary differential element of volume. Hence, we have

$$\nabla\cdot\mathbf{E} = 4\pi\rho, \tag{2.5.6}$$

at all points of the charge region. Using the expression (2.4.10) for the potential, this equation can be written[10]

$$\nabla^2\varphi = -4\pi\rho, \tag{2.5.7}$$

which is Poisson's equation. These are the differential equations of electrostatics in the absence of polarization.

In pure electrostatics there are considered to be two kinds of bodies; conductors and insulators. In a conductor (a metal) there are a large number of free charges, which are compensated at each macroscopic element of space by the bound lattice charge, which cannot move. Obviously any electric field in the conductor will cause the free charge to move and, hence, cause a current to flow. Since flowing currents are outside the realm of electrostatics, no electric field is possible inside a conductor *in electrostatics*. Then from (2.5.6) we see that $\rho = 0$ and, hence, that there is no *resultant* charge density in the interior of a conductor in electrostatics. Since from (2.4.10) we have

$$\mathbf{E} = -\nabla\varphi, \tag{2.5.8}$$

it is clear that φ is constant in a conductor in electrostatics. However, there can be and, indeed, is charge on the surface of a conductor, but the potential is constant *on the surface* as well as in the interior.

On the other hand, in an insulator (a dielectric such as quartz) there are no free charges which can move. In an insulator there is positive lattice charge and negative electronic and lattice charge. Although the positive and negative charge usually compensate each other, they need not and there may be a net charge density ρ. Although the positive and negative charges cannot have a velocity they can displace slightly relative to each other and thus produce a dipole moment (or polarization). This phenomena of polarizing is characteristic of a dielectric or insulator and accounts for its electrical behavior. Consequently, we will determine the influence on the field variable $\mathbf{E}$ and φ of a dipole moment or polarization in the next chapter.

CHAPTER 3

Electric Field Equations in Charged and Polarized Regions

3.1 The Electric Dipole Moment

We begin by considering two point charges $+q$ and $-q$ separated by a distance l as shown in Figure 10. The schematic shown in Figure 10 is a portion of the insulating body or dielectric material as shown in Figure 11, where for the present P is outside the body or charge region. The quantity

$$\mathbf{m} = q\hat{\mathbf{a}}l, \tag{3.1.1}$$

is called the electric dipole moment of the two charges $+q$ and $-q$ where $\hat{\mathbf{a}}$ is a unit vector directed from $-q$ to $+q$ as shown in Figure 10. We are particularly interested in the situation in which $q \to \infty$ $l \to 0$ while the product ql remains finite and equal to $|\mathbf{m}| = m$. Thus, we have

$$\mathbf{m} = \lim_{\substack{l \to 0 \\ q \to \infty}} ql\hat{\mathbf{a}} = m\hat{\mathbf{a}}. \tag{3.1.2}$$

We now wish to determine the potential φ at P due to a dipole $\mathbf{m}$. Clearly

$$\varphi = -\frac{q}{r_1} + \frac{q}{r_2} = q\left(-\frac{1}{r_1} + \frac{1}{r_2}\right), \tag{3.1.3}$$

but in the limit $l \to 0$, we have

$$\frac{1}{r_2} = \frac{1}{r_1} + d\left(\frac{1}{r_1}\right) = \frac{1}{r_1} + \hat{\mathbf{a}}l \cdot \nabla^Q\left(\frac{1}{r_1}\right), \tag{3.1.4}$$

where ∇^Q is used because in this operation the field point P remains fixed and the source points Q are changed. Thus, substituting (3.1.4) into (3.1.3) and employing (3.1.2), we obtain

$$\varphi = q\hat{\mathbf{a}}l \cdot \nabla^Q\left(\frac{1}{r_1}\right) = \mathbf{m} \cdot \nabla^Q\left(\frac{1}{r_1}\right). \tag{3.1.5}$$

This expression for the potential φ is for a point dipole moment $\mathbf{m}$ at Q. Clearly, we may repeat this for a number of discrete dipole moments at

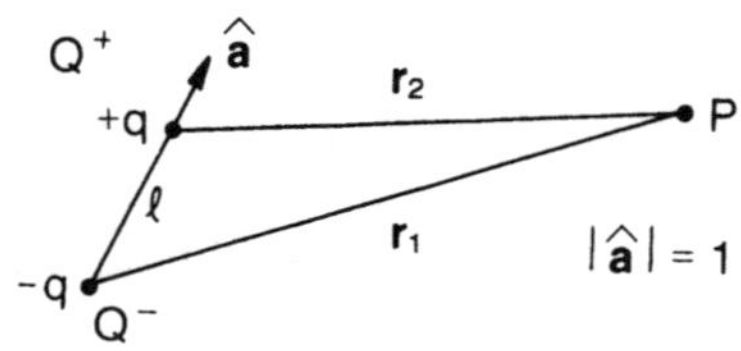

FIGURE 10. Schematic of electric dipole moment.

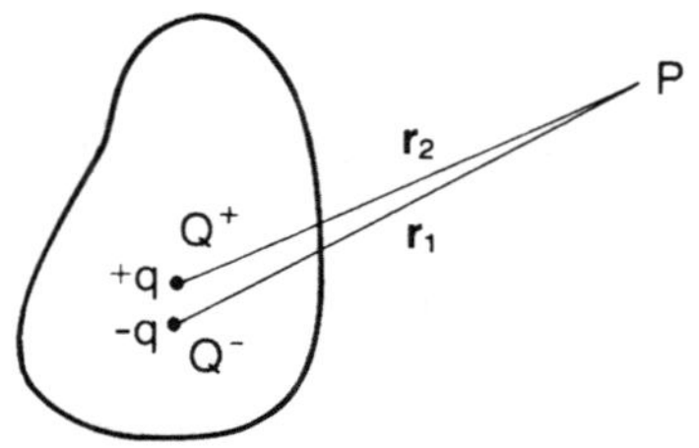

FIGURE 11. Dielectric body.

distinct points $Q^{(n)}$ just as we did for point charges near the end of Sec. 2.1 to obtain

$$\varphi(P) = \sum_{n=1}^{N} \mathbf{m}^{(n)} \cdot \nabla^{Q^{(n)}} \left(\frac{1}{r_1^n} \right). \tag{3.1.6}$$

Then by means of an argument directly analogous to the one employed at the beginning of Sec. 2.2 to define the continuous distribution of charge density ρ from a distribution of discrete point charges $q^{(n)}$, we may define a continuous distribution of electric dipole moment density $\mathbf{P}$ from the distribution of point dipole moments $\mathbf{m}^{(n)}$. In this way, just as we did in the case of charge, we replace $\sum_n \mathbf{m}^{(n)}$ at the Q points in the discrete model by $\int \mathbf{P}\, dV$ at the Q points in the continuum, and from (3.1.6) for a continuous distribution of polarization $\mathbf{P}$, we may write

$$\varphi(P) = \int_V \mathbf{P} \cdot \nabla^Q \left(\frac{1}{r_1} \right) dV, \tag{3.1.7}$$

where the field point P must be outside the polarized region. Clearly Eq. (3.1.7) can also be written in the form

$$\varphi(P) = -\int_V \mathbf{P} \cdot \nabla^P \left(\frac{1}{r_1} \right) dV, \tag{3.1.8}$$

and since the field point is presently constrained to be outside the polarized region, r_1 never becomes zero, so that we can differentiate under the integral sign and obtain

$$\mathbf{E} = -\nabla^P \varphi = \int_V \mathbf{P} \cdot \nabla^P \nabla^P \left(\frac{1}{r_1} \right) dV. \tag{3.1.9}$$

Since in this discussion we are supposing that the net charge density ρ is zero everywhere, we have

$$\nabla^P \cdot \mathbf{E} = -\nabla^P \cdot \nabla^P \varphi = 0 \tag{3.1.10}$$

outside the dielectric.

3.2 Fields Inside the Polarized Region

We now wish to determine expressions for the field and potential in the interior of the polarized region. In the case of the potential φ, it is clear from the form of the expression for φ and the fact that the dependence on r_1 is the same as for the electric field due to a charge density ρ, which we have proven is valid within the charge region, that the same expression for φ is valid within the polarized region. In the case of the electric field $\mathbf{E}$ our previous work should already tell us that this will not be the case.[11] The reason for this is that from (3.1.9), we have

$$\mathbf{E} = \int_V \mathbf{P} \cdot \nabla^P \nabla^P \left(\frac{1}{r_1} \right), \tag{3.2.1}$$

and from (2.4.6)

$$\nabla^P \left(\frac{1}{r_1} \right) = -\frac{\mathbf{r}_1}{r_1^3}, \tag{3.2.2}$$

so that with (2.2.1) we may write

$$\nabla^P \nabla^P \left(\frac{1}{r_1} \right) = -\nabla^P \frac{\mathbf{r}_1}{r_1^3} = -\mathbf{e}_j \frac{\partial}{\partial x_j} \frac{(x_i - \xi_i)}{r_1^3} \mathbf{e}_i, \tag{3.2.3}$$

and

$$\frac{\partial}{\partial x_j} \left(\frac{x_i - \xi_i}{r_1^3} \right) = \frac{\delta_{ij}}{r_1^3} - \frac{3(x_i - \xi_i)}{r_1^4} \frac{(x_j - \xi_j)}{r_1},$$

hence with (3.2.1) we have

$$E_i = \int_V \left[-\frac{P_i}{r_1^3} + 3 \frac{P_j (x_j - \xi_j)(x_i - \xi_i)}{r_1^5} \right] dV,$$

$$\mathbf{E} = \int_V \left[-\frac{\mathbf{P}}{r_1^3} + 3 \frac{\mathbf{P} \cdot \mathbf{r}_1 \mathbf{r}_1}{r_1^5} \right] dV, \tag{3.2.4}$$

which is not of the form that our demonstration has shown to be convergent when the singular point is included in the domain of integration because here $n = 3$. As a consequence we proceed in a different way in order to establish a well-defined theory.

A convenient starting point is the expression for φ in the form (3.1.7) which may be written

$$\varphi = \int_V \nabla^Q \cdot \left(\frac{\mathbf{P}}{r_1}\right) dV - \int_V \frac{\nabla^Q \cdot \mathbf{P}}{r_1} dV. \tag{3.2.5}$$

We now consider the region shown in Figure 12 in which we surround the field point P by a surface Σ enclosing a small volume V'', which we exclude from the domain of integration in the expression for φ that we use in calculating $\mathbf{E}$, and take the limit as $V'' \to 0$ as was done in Sect. 2.4. In other words we have

$$\varphi' = \int_{V'} \nabla^Q \cdot \left(\frac{\mathbf{P}}{r_1}\right) dV - \int_{V'} \frac{\nabla^Q \cdot \mathbf{P}}{r_1} dV, \tag{3.2.6}$$

and we know that $\varphi' = \varphi$ in the limit $V'' \to 0$ $(V' \to V)$ by virtue of (3.1.8) and (2.4.8). Applying the divergence theorem to (3.2.6), we obtain

$$\varphi' = \int_S \frac{\mathbf{n} \cdot \mathbf{P}}{r_1} dS + \int_\Sigma \frac{\mathbf{n} \cdot \mathbf{P}}{r_1} dS - \int_{V'} \frac{\nabla^Q \cdot \mathbf{P}}{r_1} dV, \tag{3.2.7}$$

which with (2.2.2) and (2.2.9) shows that the potential of the polarization may be considered to be the same as that due to a volume distribution of charge

$$\bar{\rho} = -\nabla^Q \cdot \mathbf{P}, \tag{3.2.8}$$

and a surface disribution of charge

$$\bar{\sigma} = \mathbf{n} \cdot \mathbf{P}, \tag{3.2.9}$$

on the surface of the polarized region S and the surface Σ excluding the domain V''. It is also clear from the expression for φ' that

$$\varphi = \lim_{V'' \to 0} \varphi', \tag{3.2.10}$$

since it is clear that the integral over Σ vanishes if $|\mathbf{P}|$ is bounded and $r_1 \to 0$. However, as we shall see this is not true for $\mathbf{E}$.

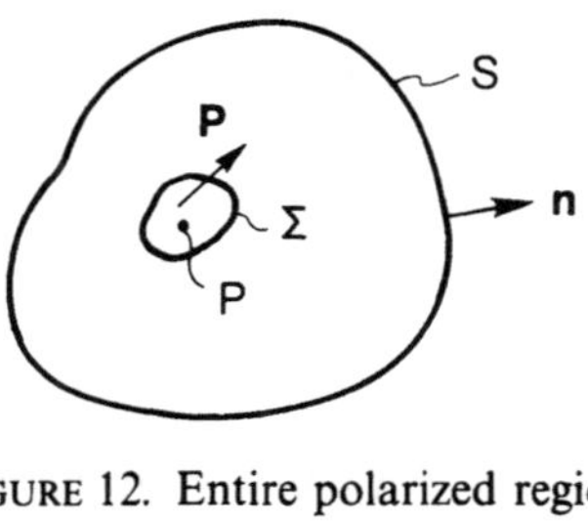

FIGURE 12. Entire polarized region.

The field $\mathbf{E}'$ is defined as

$$\mathbf{E}' = -\nabla^P \varphi, \tag{3.2.11}$$

hence from (3.2.7), we obtain

$$\mathbf{E}' = -\int_S \mathbf{n}\cdot\mathbf{P}\nabla^P\left(\frac{1}{r_1}\right)dS - \int_\Sigma \mathbf{n}\cdot\mathbf{P}\nabla^P\left(\frac{1}{r_1}\right)dS + \int_{V'} \nabla^Q\cdot\mathbf{P}\nabla^P\left(\frac{1}{r_1}\right)dV, \tag{3.2.12}$$

from which with (3.2.2) it is clear that the integral over Σ gives a finite contribution[12] as $r_1 \to 0$, and the contribution depends on the shape of Σ. Since $\Sigma \to 0$ this is a *local* field and depends on the state of the medium at the point P only. Moreover, for the development of the mechanical description of the behavior of the continuum at the point P, which is of ultimate interest to this writer, the effect of this local force field can very readily be incorporated in the definition of the local mechanical traction vector. On the other hand, if a microscopic description of the lattice is to be employed as in various aspects of solid state physics, along with the macroscopic electric field rather than the free-space field, this force field must be included and is called the Lorentz local field. In any event as far as electrostatics (or electromagnetics) is concerned the arbitrary course of simply ignoring this finite force field is adopted. This step is essential to developing a workable electrostatic (or electromagnetic) field theory.[13]

In accordance with the foregoing discussion we now define $\mathbf{E}''$ as

$$\mathbf{E}'' = -\int_S \mathbf{n}\cdot\mathbf{P}\nabla^P\left(\frac{1}{r_1}\right)dS + \int_{V'} \nabla^Q\cdot\mathbf{P}\nabla^P\left(\frac{1}{r_1}\right)dV, \tag{3.2.13}$$

so that

$$\mathbf{E}' = \mathbf{E}'' - \int_\Sigma \mathbf{n}\cdot\mathbf{P}\nabla^P\left(\frac{1}{r_1}\right)dS, \tag{3.2.14}$$

and then define the electric field $\mathbf{E}$ as

$$\mathbf{E} = \lim_{V''\to 0} \mathbf{E}'', \quad V' \to V, \tag{3.2.15}$$

from which, in view of the preceding discussion, it is clear that the electric field does not represent the force per unit charge in a polarized region. In order to obtain the force per unit charge this local field must be included. This fact affords no difficulty in a complete continuum description because the mechanical force effect of this local field can readily be incorporated in the mechanical traction vector. In any event this is the field $\mathbf{E}$ that appears from this point on in electrostatic (or electromagnetic) theory. Since it is clear that $\varphi'' = \varphi' = \varphi$ in the limit $V'' \to 0$, we still have

$$\mathbf{E} = -\nabla^P \varphi, \tag{3.2.16}$$

within the polarized region.

It should, of course, have been understood that in the foregoing treatment of polarization the net charge density ρ has been assumed to vanish. If there is both charge density ρ and polarization $\mathbf{P}$, the resultant potential φ and electric field $\mathbf{E}$ may be obtained by superposition. In other words in the presence of charge and polarization, from (2.2.10), (2.4.9) and (3.2.13) and (3.2.15) the expression for $\mathbf{E}$ *inside* or outside the polarized region may be written in the form

$$\mathbf{E} = -\int_S \sigma \nabla^P\left(\frac{1}{r_1}\right) dS - \int_V \rho \nabla^P\left(\frac{1}{r_1}\right) dV - \int_S \mathbf{n}\cdot\mathbf{P}\nabla^P\left(\frac{1}{r_1}\right) dS + \int_V \nabla^Q\cdot\mathbf{P}\nabla^P\left(\frac{1}{r_1}\right) dV. \tag{3.2.17}$$

There are other forms, but this form is particularly useful for our purposes because the dependence on the field point of each term is of the form

$$\nabla^P\left(\frac{1}{r_1}\right) = -\frac{\mathbf{r}_1}{r_1^3}. \tag{3.2.18}$$

It is perfectly clear from our previous work in Sec. 2.4 that if ρ and $\nabla^Q\cdot\mathbf{P}$ are bounded and integrable φ and $\mathbf{E}$ are continuous, and that if ρ and $\nabla^Q\cdot\mathbf{P}$ are continuous along with their first spatial derivatives, the first derivatives of $\mathbf{E}$ and second derivatives of φ exist and are continuous at interior points.

3.3 Field Equation of Electrostatics

Consider the entire charged and polarized region, which is enclosed within S_0 shown in Figure 13, and consider any surface S within which ρ, $\nabla^Q\cdot\mathbf{P}$ and their gradients are continuous. Integrate the normal component of $\mathbf{E}$ over S to from the scalar

$$\int_S \mathbf{n}\cdot\mathbf{E}\, dS, \tag{3.3.1}$$

which, since $\mathbf{E}$ is the electric field due to all charge and polarization contained within S_0, with (3.2.17) and (3.2.18) enables us to write

$$\int_S \mathbf{n}\cdot\mathbf{E}\, dS = \int_S \mathbf{n}\cdot\left[\int_{V_0} (\rho - \nabla^Q\cdot\mathbf{P})\frac{\mathbf{r}_1}{r_1^3} dV + \int_{S_0} (\sigma + \mathbf{n}\cdot\mathbf{P})\frac{\mathbf{r}_1}{r_1^3} dS_0\right] dS \tag{3.3.2}$$

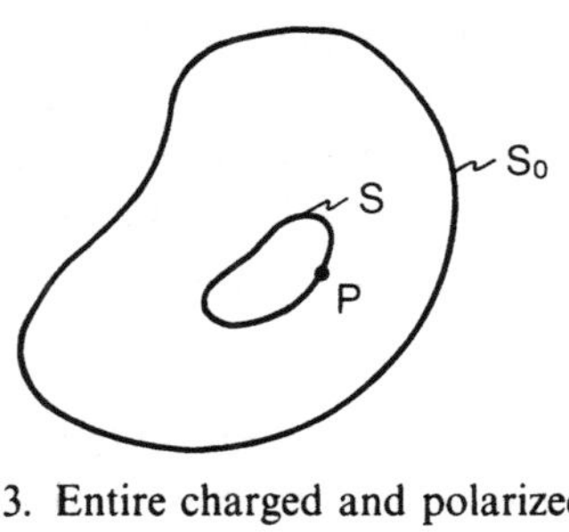

FIGURE 13. Entire charged and polarized region.

Interchanging the order of integration, we obtain

$$\int_S \mathbf{n}\cdot\mathbf{E}\,dS = \int_{V_0} (\rho - \nabla^Q\cdot\mathbf{P}) \int_S \frac{\mathbf{n}\cdot r_1}{r_1^3}\,dS\,dV + \int_{S_0} (\sigma + \mathbf{n}\cdot\mathbf{P}) \int_S \frac{\mathbf{n}\cdot\mathbf{r}_1}{r_1^3}\,dS\,dS_0. \tag{3.3.3}$$

Integrating the right-hand side of (3.3.3) over the closed surface S and employing (2.3.7)–(2.3.9), we obtain

$$\int_S \mathbf{n}\cdot\mathbf{E}\,\mathrm{d}S = 4\pi \int_V (\rho - \nabla^Q\cdot\mathbf{P})\,dV, \tag{3.3.4}$$

since, as shown in Sec. 2.3, for a closed surface S, the solid angle for any point outside S vanishes and for any point inside is 4π, and all of S_0 and $V_0 - V$ are outside S. Since the integration variables are dummy variables and we restrict ρ and $\nabla^Q\cdot\mathbf{P}$ to satisfy the conditions of the divergence theorem, we obtain

$$\int_V (\nabla\cdot\mathbf{E} + 4\pi\nabla\cdot\mathbf{P} - 4\pi\rho)\,dV = 0, \tag{3.3.5}$$

for any volume V, and in particular an arbitrary differential element of volume. Hence, we obtain

$$\nabla\cdot(\mathbf{E} + 4\pi P) = 4\pi\rho, \tag{3.3.6}$$

as the differential equation of electrostatics valid for insulators and conductors alike.

The vector quantity $\mathbf{E} + 4\pi\mathbf{P}$ is a very important electrical quantity called the electric displacement and is given the symbol $\mathbf{D}$. Thus, we define

$$\mathbf{D} = \mathbf{E} + 4\pi\mathbf{P}, \tag{3.3.7}$$

in Gaussian (or c.g.s.) units, which enables us to write the differential equations of electrostatics in the form

$$\nabla\cdot\mathbf{D} = 4\pi\rho, \tag{3.3.8}$$

$$\mathbf{D} = \mathbf{E} + 4\pi\mathbf{P}, \tag{3.3.9}$$

$$\mathbf{E} = -\nabla\varphi. \tag{3.3.10}$$

These equations in this form are quite general but cannot be solved unless there is either a relation between $\mathbf{P}$ and $\mathbf{E}$ or $\mathbf{P}$ is given.

3.4 Linearized Electrostatics

In ordinary elementary electrostatics a linear relation between $\mathbf{P}$ and $\mathbf{E}$ is assumed, i.e.,

$$\mathbf{P} = \chi\mathbf{E}, \tag{3.4.1}$$

for an isotropic material and

$$P_i = \chi_{ij} E_j, \tag{3.4.2}$$

for an anisotropic material, where χ_{ij} is not yet symmetric. The symbol χ is called the electric susceptibility and χ_{ij} the susceptibility tensor. It is of second rank. When these linear relationships exist we have

$$D_i = E_i + 4\pi P_i = (\delta_{ij} + 4\pi\chi_{ij})E_j = \varepsilon_{ij}E_j, \tag{3.4.3}$$

where

$$\varepsilon_{ij} = \delta_{ij} + 4\pi\chi_{ij}, \tag{3.4.4}$$

is the dielectric tensor. For an isotropic material we have

$$\varepsilon_{ij} = \delta_{ij}(1 + 4\pi\chi) = \varepsilon\delta_{ij}, \tag{3.4.5}$$

so that we may write

$$\mathbf{D} = \varepsilon\mathbf{E}, \tag{3.4.6}$$

where ε is called the dielectric constant. Then in an isotropic linear electrostatic problem we have

$$\nabla^2\varphi = -\frac{4\pi\rho}{\varepsilon}, \tag{3.4.7}$$

as the scalar differential equation valid in each region with different ε and ρ. This is a Poisson equation and to this we have to adjoin the boundary conditions. However, before we do this we observe that in actuality we have no particular interest in the linear constitutive equations we have written in this section, only in the development and the equations up through Sec. 3.3. Hence, from our point of view we need the boundary conditions that go along with (3.3.8)–(3.3.10).

3.5 Boundary Conditions of Electrostatics

The differential equations of electrostatics are given by (3.3.8)–(3.3.10), and we note that (3.3.10) can be written in the equivalent form

$$\nabla \times \mathbf{E} = 0. \tag{3.5.1}$$

Now, in any continuum field theory the truly fundamental equations must always be integral statements, from which the governing differential equations may be derived when suitable differentiability and continuity conditions are satisfied. The integral statements are more general because they are *postulated* (taken) to hold even where the differential equations do not, as across a surface of discontinuity. In fact the integral forms may be used to find the boundary (or continuity) conditions across surfaces of discontinuity. The integral forms, corresponding to (3.3.8) and (3.5.1), respectively, are

$$\int_S \mathbf{n}\cdot\mathbf{D}\,dS = 4\pi \int_V \rho\,dV, \tag{3.5.2}$$

$$\oint_C \mathbf{E}\cdot d\mathbf{r} = 0, \tag{3.5.3}$$

where S is *any* closed surface enclosing a volume V and C is *any* closed curve. These are the appropriate integral forms because, if **D** and **E** are sufficiently differentiable and continuous for the divergence theorem and Stokes theorem to be valid, they yield (3.3.8) and (3.5.1) or (3.3.10), respectively. However, as already mentioned, the integral forms are valid even when the differential forms are not.

Consider a surface of discontinuity across which **D** and **E** may be discontinuous. Note that at this surface ρ may be singular in such a way that[14]

$$\lim_{\substack{\rho\to\infty \\ \Delta l\to 0}} \rho\Delta V = \rho\Delta l\Delta a = \sigma\Delta a, \tag{3.5.4}$$

where σ is the finite limit

$$\sigma = \lim_{\substack{\rho\to\infty \\ \Delta l\to 0}} \rho\Delta l, \tag{3.5.5}$$

and is the surface charge density. To determine the conditions on **D** consider the pill-box region shown in Figure 14 and apply Eq. (3.5.2) as $\mathscr{S}^+$ and $\mathscr{S}^-$ both approach s while the volumes $\mathfrak{R}^+$ and $\mathfrak{R}^-$ vanish so that the area of S remains finite in the limit. Thus, we have

$$\lim_{\substack{\mathfrak{R}^+\to 0 \\ \mathfrak{R}^+ - 0}} \int_{\mathscr{S}^+} \mathbf{v}\cdot\mathbf{D}^+\,dS + \int_{\mathscr{S}^-} \mathbf{v}\cdot\mathbf{D}^-\,dS = 4\pi\left[\int_{\mathfrak{R}^+} \rho^+\,dV + \int_{\mathfrak{R}^-} \rho^-\,dV\right], \tag{3.5.6}$$

and if ρ remained bounded we could obtain the result directly with a zero right-hand side. Since ρ may diverge at s, we write in the limit

$$\int_S (\mathbf{n}\cdot\mathbf{D}^+ - \mathbf{n}\cdot\mathbf{D}^-)\,dS = \lim_{\substack{t^+\to 0 \\ t^-\to 0 \\ \rho^+\to\infty \\ \rho^-\to\infty}} 4\pi \int_S (\rho^+ t^+ + \rho^- t^-)\,dS, \tag{3.5.7}$$

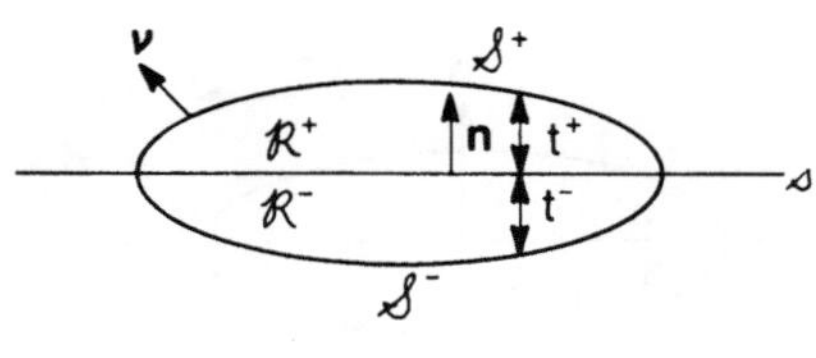

FIGURE 14. Pill-box region.

and since in the limit we have

$$\sigma = \lim_{\substack{t^+ \to 0 \\ t^- \to 0 \\ \rho^+ \to \infty \\ \rho^- \to \infty}} (\rho^+ t^+ + \rho^- t^-), \tag{3.5.8}$$

we may write

$$\int_S [\mathbf{n}\cdot(\mathbf{D}^+ - \mathbf{D}^-) - 4\pi\sigma]\,dS = 0, \tag{3.5.9}$$

for *any* surface area A on S, including in particular *any* arbitrary infinitesimal area. Hence, we have

$$\mathbf{n}\cdot(\mathbf{D}^+ - \mathbf{D}^-) = 4\pi\sigma, \tag{3.5.10}$$

as the condition on $\mathbf{D}$ across a discontinuity surface s. This condition is called a jump condition and it is conventional to write this in the form

$$\mathbf{n}\cdot[\mathbf{D}] = 4\pi\sigma, \tag{3.5.11}$$

where

$$[\mathbf{D}] \equiv \mathbf{D}^+ - \mathbf{D}^-. \tag{3.5.12}$$

Note that our jump condition says that the normal component of $\mathbf{D}$ is continuous across any surface on which there is no surface charge, and if there is a surface charge it is related to the discontinuity in the normal component of $\mathbf{D}$ across the surface through Eq. (3.5.11). It should be noted that in just about all practical situations, there is no charge density ρ in an insulator as well as in a conductor, and as a consequence, there is no surface charge at a dielectric–dielectric interface or at a conductor (metal)–conductor interface. Clearly then, the normal component of $\mathbf{D}$ is continuous across these interfaces. However, there almost always is a surface charge and, hence, a discontinuity in $\mathbf{n}\cdot\mathbf{D}$ across a dielectric–conductor interface.

To determine the condition on $\mathbf{E}$ across a discontinuity surface S consider the circulation of $\mathbf{E}$ around the curve $C^+ + C^- = C$ shown surrounding S in Figure 15 and take the limit as s^+ and s^- both approach s while the areas $\mathscr{A}^+$ and $\mathscr{A}^-$ vanish so that the length l of s remains finite in the limit. Accordingly we write

$$\lim_{\substack{\mathscr{A}^+ \to 0 \\ \mathscr{A}^- \to 0}} \int_{C^+} \mathbf{E}^+\cdot d\mathbf{r} + \int_{C^-} \mathbf{E}^-\cdot d\mathbf{r} = 0, \tag{3.5.13}$$

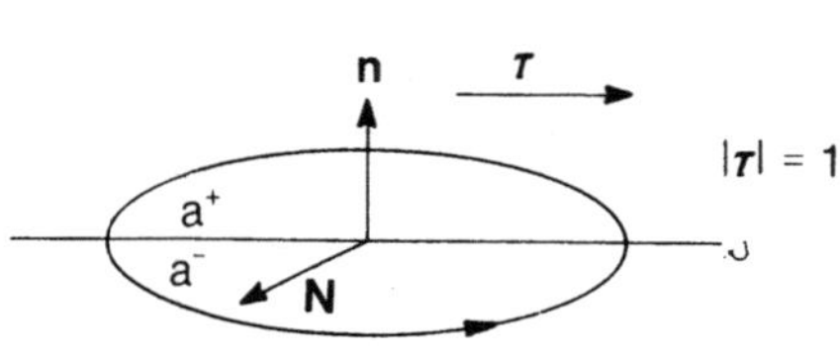

FIGURE 15. Circulation around surface of discontinuity.

which may be written

$$-\int_l \boldsymbol{\tau}\cdot(\mathbf{E}^+ - \mathbf{E}^-)ds = 0, \tag{3.5.14}$$

where $\boldsymbol{\tau}$ denotes an *arbitrary* unit vector lying in s. Clearly we can write

$$\boldsymbol{\tau} = \mathbf{n} \times \mathbf{N}, \tag{3.5.15}$$

where $\mathbf{N}$ denotes a unit vector normal to the curve C and lying in s. Then we have

$$\int_l (\mathbf{E}^+ - \mathbf{E}^-)\cdot\mathbf{N} \times \mathbf{n}\,ds = 0, \tag{3.5.16}$$

and by a well-known vector identity we may write

$$\int_l \mathbf{N}\cdot\mathbf{n} \times (\mathbf{E}^+ - \mathbf{E}^-)ds = 0 \tag{3.5.17}$$

for *any* line element l on s and any accompanying $\mathbf{N}$, including in particular *any* arbitrary infinitesimal l and the associated $\mathbf{N}$. Hence we have

$$\mathbf{n} \times (\mathbf{E}^+ - \mathbf{E}^-) = 0, \tag{3.5.18}$$

as the condition on $\mathbf{E}$ across a discontinuity surface S. Using our previous convention for denoting jumps, we have

$$\mathbf{n} \times [\mathbf{E}] = 0, \tag{3.5.19}$$

which says that the tangential component of $\mathbf{E}$ is continuous across any surface of discontinuity. Moreover, since in electrostatics $\mathbf{E} = -\nabla\varphi$, the tangential components of $\mathbf{E}$ are determined by the surface gradient of φ, which is determined by the values of φ at each point of the surface, and we have the equivalent condition $[\varphi] = K$ across S, where K is at most a constant, which is physically irrelevant. Hence across a discontinuity surface S, we have

$$[\varphi] = 0, \tag{3.5.20}$$

as the condition on φ.

An important consequence of the foregoing is that φ is continuous and differentiable at points *on* the charged surfaces if σ remains bounded. The surface charge σ will be assumed to be bounded throughout this treatment. This excludes the possibility of having a distribution of dipole moments per unit surface area, which is commonly called a double layer.[15] As might be expected the boundary condition, i.e., (3.5.19) or (3.5.20), would change drastically under these circumstances. However, we exclude this circumstance as unphysical inasmuch as it will not occur in any physical situation encountered in this treatment.

CHAPTER 4

Forces and Torques Exerted by the Electric Field on Charged and Polarized Matter

4.1 Electric Body Forces

It is quite clear from the material in Chapter 2 that by definition the force exerted by the electric field on a differential charged volume $\rho\, dV$ is given by

$$d\mathbf{F} = \mathbf{E}\rho\, dV. \tag{4.1.1}$$

If no polarization is present at the point in question this body force represents the total electric force exerted on $\rho\, dV$. However, if polarization is present, as we already know, a portion of the electric force exerted on $\rho\, dV$ is due to a local surface effect, which depends on the shape of the volume element, and in a continuum theory can readily be included as a portion of the mechanical traction vector. Nevertheless, $\mathbf{E}\rho\, dV$ is still the body force exerted by the electric field vector $\mathbf{E}$ on ρdV, since the local electric force, which depends on the shape of the surface surrounding the volume element, will be included as part of the mechanical traction.

Now, we wish to know what is the body force exerted by the electric field $\mathbf{E}$ on polarization $\mathbf{P}$ at a region dV. To this end we consider the electric field $\mathbf{E}$ (not the force on a unit point charge) in the neighborhood of two point charges $+q$ and $-q$ separated by the distance $\hat{\mathbf{a}}l$, as shown in Figure 16, just as we did in the definition of polarization, and determine the force exerted by $\mathbf{E}$ on the two charges and then take the limit of the expression as $l \to 0$ and $q \to \infty$ while the product $lq \to m$. From Figure 16 it is clear that the force is given by

$$\mathbf{f} = -q\mathbf{E}^- + q\mathbf{E}^+ = q(-\mathbf{E}^- + \mathbf{E}^+), \tag{4.1.2}$$

and in the limit $l \to 0$, we may write

$$\mathbf{E}^+ = \mathbf{E}^- + l\hat{\mathbf{a}}\cdot\nabla\mathbf{E}^-. \tag{4.1.3}$$

Hence in the limit $q \to \infty$ we have

$$\mathbf{f} = lq\hat{\mathbf{a}}\cdot\nabla^Q\mathbf{E}^{Q^-} = \mathbf{m}\cdot\nabla^Q\mathbf{E}^{Q^-}. \tag{4.1.4}$$

Clearly, we may repeat this for a number of discrete dipole moments at

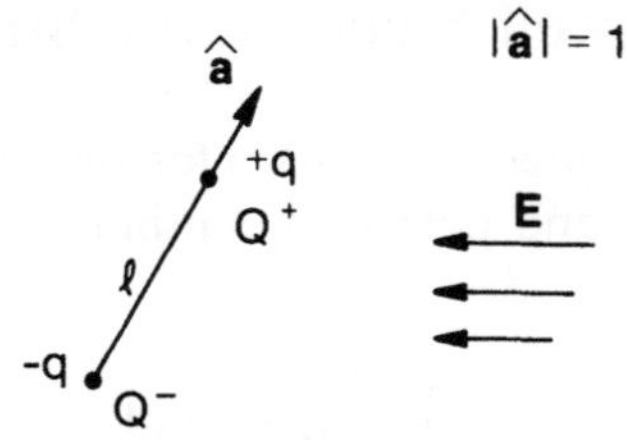

FIGURE 16. Force exerted by electric field on a point dipole.

distinct points (n) just as we did in Sec. 3.1 to obtain

$$\mathbf{f} = \sum_{n=1}^{N} \mathbf{m}^{(n)} \cdot \nabla^{(n)} \mathbf{E}^{(n)-}. \tag{4.1.5}$$

Then we note that for the microscopic region of interest, which enables us to define the macroscopic polarization density $\mathbf{P}$, the gradient of the macroscopic electric field has one value. Consequently, for a region of that size Eq. (4.1.5) may be written in the form

$$\mathbf{f} = \sum_{n=1}^{M} \mathbf{m}^{(n)} \cdot \nabla \mathbf{E}. \tag{4.1.6}$$

Then by means of reasoning exactly the same as that leading from (3.1.6) to (3.1.7), we may write

$$d\mathbf{F} = \mathbf{P} \cdot \nabla \mathbf{E}\, dV. \tag{4.1.7}$$

Clearly then, from (4.1.1) and (4.1.7) the expression for the force exerted by the electric field on continuous distributions of charge density ρ and polarization $\mathbf{P}$ in a differential element of volume dV may be written in the form

$$d\mathbf{F} = (\rho \mathbf{E} + \mathbf{P} \cdot \nabla \mathbf{E}) dV. \tag{4.1.8}$$

If we integrate (4.1.8) over some portion of or an entire body, we obtain

$$\mathbf{F} = \int_V (\rho \mathbf{E} + \mathbf{P} \cdot \nabla \mathbf{E}) dV, \tag{4.1.9}$$

where $\mathbf{F}$ is the force exerted by $\mathbf{E}$ on either some portion of or an entire body, and we note that $\mathbf{E}$ in (4.1.9) arises from sources both inside and outside the region denoted V. Since it is conventional to write

$$\mathbf{F} = \int_V \mathbf{f}\, dV, \tag{4.1.10}$$

where $\mathbf{f}$ is the body force density or body force per unit volume, we have shown that the expression for $\mathbf{f}$ exerted by $\mathbf{E}$ on charged and polarized material continua may be written in the form

$$\mathbf{f} = \rho \mathbf{E} + \mathbf{P} \cdot \nabla \mathbf{E}. \tag{4.1.11}$$

4.2 Maxwell Electrostatic Stress Tensor

Now that we have the expression (4.1.11) for the body force exerted by the electric field on charged and polarized continua, we substitute from (3.3.8) into (4.1.11) to obtain

$$\mathbf{f}=\frac{\nabla\cdot\mathbf{D}}{4\pi}\mathbf{E}+\mathbf{P}\cdot\nabla\mathbf{E}, \tag{4.2.1a}$$

or in Cartesian component form

$$f_j=\frac{1}{4\pi}\frac{\partial D_k}{\partial x_k}E_j+P_k\frac{\partial E_j}{\partial x_k}, \tag{4.2.1b}$$

which shows that the electric body force term is automatically nonlinear in the electric field variables, and hence, we already see that any coupled electromechanical system is already inherently nonlinear. Now, since we have th field description and the P or Q point makes no difference we can use the comma notation for differentiation. Accordingly, we write (4.2.1b) in the form

$$f_j=\frac{1}{4\pi}D_{k,k}E_j+P_kE_{j,k}, \tag{4.2.2}$$

which with (3.3.9) may be written

$$f_j=\frac{1}{4\pi}D_{k,k}E_j+\left(\frac{D_k-E_k}{4\pi}\right)E_{j,k}, \tag{4.2.3}$$

or

$$f_j=\frac{1}{4\pi}(D_kE_j)_{,k}-\frac{1}{4\pi}E_kE_{j,k}. \tag{4.2.4}$$

By virtue of (3.3.10) or (3.5.1), we have

$$E_{j,k}=E_{k,j}, \tag{4.2.5}$$

which with (4.2.4) enables us to write

$$f_j=\frac{1}{4\pi}[D_kE_j-\tfrac{1}{2}E_lE_l\delta_{jk}]_{,k}, \tag{4.2.6}$$

and this last expression within the brackets is clearly a second rank tensor and is called the Maxwell electrostatic strss tensor. Hence, we may write

$$T^E_{kj}=\frac{1}{4\pi}[D_kE_j-\tfrac{1}{2}E_lE_l\delta_{kj}], \tag{4.2.7}$$

for the Maxwell electrostatic stress tensor, which enables (4.2.6) to be written in the form

$$f_j=T^E_{kj,k}, \tag{4.2.8}$$

or in invariant dyadic form (4.2.7) may be written

$$\mathbf{T}^E = \frac{1}{4\pi}[\mathbf{DE} - \tfrac{1}{2}\mathbf{E}\cdot\mathbf{EI}], \tag{4.2.9}$$

and (4.2.8) may be written

$$\mathbf{f} = \nabla\cdot\mathbf{T}^E. \tag{4.2.10}$$

Consequently, the force $\mathbf{F}$ exerted by $\mathbf{E}$ on a volume V of charged and polarized matter may be written

$$\mathbf{F} = \int_V \mathbf{f}\,dV = \int_V \nabla\cdot\mathbf{T}^E\,dV, \tag{4.2.11}$$

and with the aid of the divergence theorem we obtain

$$\mathbf{F} = \int_S \mathbf{n}\cdot\mathbf{T}^E\,dS. \tag{4.2.12}$$

Thus we have shown that the force exerted by the electric field $\mathbf{E}$ on an element of matter may be represented as an integral of the Maxwell electrostatic traction force over the surface S bounding V. This is very important because, as we already know, integral statements are more fundamental than differential statements because they can exist when the differential statements do not. This is indeed the case here, where we take (postulate that) the Maxwell electrostatic stress tensor $\mathbf{T}^E$ to represent the mechanical force effect of $\mathbf{E}$ as a force acting over the surface of an arbitrary region, in general; i.e., even when the field vectors are discontinuous and $\nabla\cdot\mathbf{T}^E$ does not exist. Of course, when the field vectors are suitably differentiable, $\nabla\cdot\mathbf{T}^E$ exists and we obtain (4.2.10) from (4.2.12). However, as we already know, the field vectors may not be continuous at material surfaces of discontinuity, and at such places $\mathbf{T}^E$ must be used. Thus, it is quite clear that the expression for $\mathbf{T}^E$ will be very important in its effect on the mechanical boundary conditions at surfaces of discontinuity.[16]

At this point it is purposeful to note that if an electric field $\mathbf{E}$ exists at points of free-space, the associated Maxwell electrostatic stress tensor $\mathbf{T}^E$ exists. Both fields exist at the point in question independent of the location and values of the sources which generated $\mathbf{E}$, which may not even be known. In particular, since $\mathbf{P} = 0$ at such points we have

$$T^E_{kj} = \frac{1}{4\pi}[E_k E_j - \tfrac{1}{2}E_l E_l \delta_{kj}], \tag{4.2.13}$$

and in particular if

$$E_k = E_1 \delta_{1k}, \tag{4.2.14}$$

we have

$$T^E_{kj} = \frac{1}{4\pi}[E_1 \delta_{1k} E_1 \delta_{1j} - \tfrac{1}{2}E_1^2 \delta_{kj}], \tag{4.2.15}$$

which shows that in this case there is an electrical tension

$$T_{11} = E_1^2/8\pi, \tag{4.2.16}$$

and two electrical compressions

$$T_{22} = T_{33} = -E_1^2/8\pi, \tag{4.2.17}$$

in free space.

4.3 Electric Torques

At this point it should be noted that in general $\mathbf{T}^E$ is not symmetric. In fact, from (4.2.9) and (3.3.9) we see that its antisymmetric part is given by

$$(\mathbf{T}^E)^A = \tfrac{1}{2}(\mathbf{PE} - \mathbf{EP}), \tag{4.3.1}$$

and $(\mathbf{T}^E)^A$ will not be zero if $\mathbf{P}$ and $\mathbf{E}$ are not colinear. This means in essence that the electric field $\mathbf{E}$ can exert a couple on the polarization $\mathbf{P}$, and that the Cauchy mechanical stress tensor cannot be expected to be symmetric in general when coupling with the electric field exists.

Let us now obtain the expression for the couple exerted by $\mathbf{E}$ on $\mathbf{P}$ directly from point charge considerations. To this end we again consider the diagram shown in Figure 16 and evaluate the couple by taking moments about the − point and then taking the usual limit. The expression for a couple $\mathbf{C}$ about 0 exerted by a force $\mathbf{g}$ located a distance $\mathbf{r}$ from 0 is

$$\mathbf{C} = \mathbf{r} \times \mathbf{g}. \tag{4.3.2}$$

Applying (4.3.2) to Figure 16 in the manner stated, we obtain

$$\mathbf{C} = l\hat{\mathbf{a}} \times q\mathbf{E} = ql\hat{\mathbf{a}} \times \mathbf{E} = \mathbf{m} \times \mathbf{E}. \tag{4.3.3}$$

Then by means of an argument similar to the one employed in going from (4.1.4) to (4.1.7), we obtain

$$\mathbf{C} = \int_V \mathbf{P} \times \mathbf{E}\, dV = \int_V \mathbf{c}\, dV. \tag{4.3.4}$$

In component form we have

$$c_i = e_{ijk} P_j E_k, \tag{4.3.5}$$

where c_i is an axial vector representing the torque density and e_{ijk} is the skew symmetric tensor. It can be shown in the usual way that this couple is equivalent to the antisymmetric portion of the Maxwell electrostatic stress tensor, i.e., by writing the antisymmetric stress tensor representation of the axial vector thus

$$\begin{aligned}(T_{lm}^E)^A &= \tfrac{1}{2} e_{ilm} c_i = \tfrac{1}{2} e_{ilm} e_{ijk} P_j E_k \\ &= \tfrac{1}{2}(\delta_{lj}\delta_{mk} - \delta_{lk}\delta_{mj}) P_j E_k \\ &= \tfrac{1}{2}(P_l E_m - E_l P_m),\end{aligned} \tag{4.3.6}$$

which is the component form of (4.3.1).

CHAPTER 5

Electrostatic Energy

5.1 Energy Resulting from Distributions of Charge

This brings us to a consideration of electrostatic energy. We begin by proceeding in the usual way, i.e., by considering the energy associated with a collection of point charges $q^{(n)}$. We first bring the first charge $q^{(1)}$ to a position and no work is done. Then we bring another charge from ∞ to its position and we determine the potential energy due to it as the negative of the work done by the field (or the work done on the field) in bringing it from ∞ to its position in the presence of 1. Continuing, then we bring a third charge in from ∞ to its position in the presence of 1 and 2 and then another and so on to $n = N$. Clearly then using Eq. (2.1.9) for each point charge, we may write

$$u_1 = 0$$

$$u_2 = \frac{q^{(1)}q^{(2)}}{r^{12}},$$

$$u_3 = \frac{q^{(1)}q^{(3)}}{r^{13}} + \frac{q^{(2)}q^{(3)}}{r^{23}},$$

$$u_4 = \cdots$$

$$u_5 = \cdots \text{and so on to } u_N, \qquad (5.1.2)$$

where u_n represents the energy of the nth charge at its position. Then the total field energy $\mathcal{U}$ may be written

$$\mathcal{U} = u_1 + u_2 + u_3 + \cdots + u_N, \qquad (5.1.2)$$

which clearly may be written in the entirely equivalent form

$$\mathcal{U} = \tfrac{1}{2}q^{(1)}\left(\frac{q^{(2)}}{r^{12}} + \frac{q^{(3)}}{r^{13}} + \cdots \frac{q^{(N)}}{r^{1N}}\right) + \tfrac{1}{2}q^{(2)}\left(\frac{q^{(1)}}{r^{21}} + \frac{q^{(3)}}{r^{23}} + \cdots \frac{q^{(N)}}{r^{2N}}\right) + \cdots$$
$$+ \tfrac{1}{2}q^{(N)}\left(\frac{q^{(1)}}{r^{N1}} + \frac{q^{(2)}}{r^{N2}} + \cdots + \frac{q^{(N-1)}}{r^{N(N-1)}}\right). \qquad (5.1.3)$$

It is clear from (2.1.14) that this may be written in the form

$$\mathscr{U} = \tfrac{1}{2} \sum_{n=1}^{N} q^{(n)} \varphi^{(n)}, \tag{5.1.4}$$

where $\varphi^{(n)}$ is the electric potential at the position of the nth charge due to all the other charges.

At this point before proceeding we note a very important fact concerning a conceptual difference between the discrete point charge description (or approach) and the continuous distributed charge density (or field theoretic) approach (description). That difference is a consequence of the fact that in the point charge description only the field energy of the entire configuration of point charges is meaningful, i.e., the field energy density at a localized region of the space has no meaning. On the other hand, the situation is quite different in the field theoretic approach, in that then the energy density of *any* localized portion of the space is meaningful and important, in fact it has to be in order to have a sensible field theory. Two other differences between the descriptions are:

1. In the point charge description, the differential equations are meaningful only outside all point charges while in the field approach, the differential equations exist and are valid within the regions of charge and polarization density, and in fact the differential equations relate the field variables to the charge and polarization densities. In other words the point charges are foreigners to the field theory and the charge and polarization densities are not.
2. In the point charge description the concept of a Maxwell electrostatic stress tensor defined at each point of space is meaningless and of no value whereas in the field description it is meaningful and has great value.

We will now apply the formula (5.1.4) for the field energy to the problem of a cavity in a metal, where the cavity is a vacuum. A diagram of the configuration is shown in Figure 17. Since there can be no volumetric charge density in either the vacuum or the metal, it is clear that we can have only surface charge density on the metal-vacuum interface. Consequently, when we apply Eq. (5.1.4) for electrostatic field energy, and employ the, by now, well-known argument to go from the discrete model to the continuous

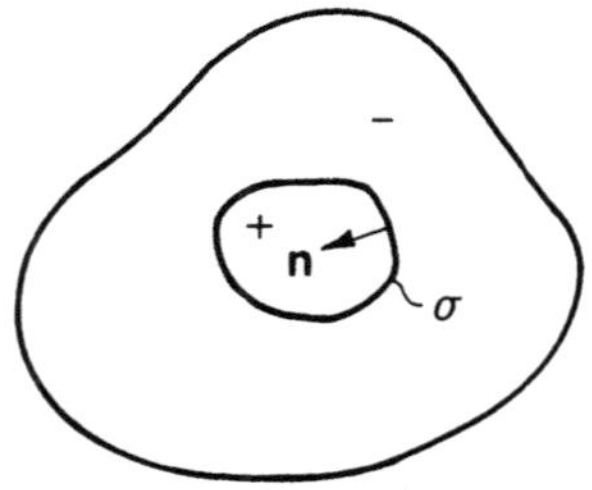

FIGURE 17. Interior vacuum in a metal.

distribution, we obtain

$$\mathscr{U} = \frac{1}{2} \int_S \sigma \varphi \, dS. \tag{5.1.5}$$

Moreover, since in this case no polarization exists anywhere, from (3.5.10) we have

$$\mathbf{n} \cdot (\mathbf{E}^+ - \mathbf{E}^-) = 4\pi\sigma, \tag{5.1.6}$$

and since the electric field in the metal $\mathbf{E}^+ \equiv 0$, we find

$$\sigma = -\frac{n_i E_i^-}{4\pi} = +\frac{n_i \varphi_{,i}}{4\pi}, \tag{5.1.7}$$

where φ is the electric potential in the vacuum. The substitution of (5.1.7) into (5.1.5) yields

$$\mathscr{U} = \frac{1}{8\pi} \int_S n_i \varphi_{,i} \varphi \, dS, \tag{5.1.8}$$

which, with the aid of the divergence theorem, yields

$$\mathscr{U} = \frac{1}{8\pi} \int_V (\varphi_{,i}\varphi)_{,i} \, dV = \frac{1}{8\pi} \int_V [\varphi_{,i}\varphi_{,i} + \varphi\varphi_{,ii}] \, dV. \tag{5.1.9}$$

Furthermore, since $\rho = 0$ in the cavity, with (2.5.7) and (2.4.10) in (5.1.9), we finally obtain

$$\mathscr{U} = \frac{1}{8\pi} \int_V E_i E_i \, dV, \tag{5.1.10}$$

which shows that in this case the surface integral of electric field energy due to surface charge is mathematically equivalent to a volume energy density in the vacuum of

$$\mathbf{E} \cdot \mathbf{E} / 8\pi \tag{5.1.11}$$

in Gaussian units.

Now, as you will recall, we have consistently used the point charge approach because of its physical clarity simply to motivate the field theory, and once the field theory has been obtained we have abandoned the point charge approach and generalized the field theory in a physically sensible way. We proceed with the same approach here also, but now in a somewhat bolder manner. We here assert (postulate) that the mathematical equivalence between the field energy due to surface charge and the volumetric electric field energy density we have just obtained is more than a mathematical equivalence, i.e., is a physical fact. This means that we associate with each volume element of free-space (or vacuum) in which an electric field exists an electric field energy density of $E^2/8\pi$. This means that in a situation such as the one we are considering, if we find an electric field $\mathbf{E}$ at a point, and we don't fully know how it got there, i.e., we can't see out to the surface charge because it is too far away or whatever, we have a physical energy density $E^2/8\pi$ at that point. This is a very important and significant departure from the point charge point of view, and is crucial to the development of the field theory.

5.2 Energy Resulting from Distributions of Charge and Polarization

We have so far said nothing about the stored energy due to the presence of dipole moments (or polarization). In order to discuss the energy of polarization we consider the usual diagram shown in Figure 10 and write the expression

$$u = -q\varphi^- + q\varphi^+ = q(-\varphi^- + \varphi^+), \tag{5.2.1}$$

for the energy due to the dipole in the presence of a field (or potential) due to *external* sources. Since as $l \to 0$

$$\varphi^+ = \varphi^- + l\hat{\mathbf{a}}\cdot\nabla\varphi^-, \tag{5.2.2}$$

we find

$$u = ql\hat{\mathbf{a}}\nabla\varphi = -ql\hat{\mathbf{a}}\cdot\mathbf{E} = -\mathbf{m}\cdot\mathbf{E}, \tag{5.2.3}$$

as the expression for the energy of a dipole moment **m** in an *external* field **E**. We thus see, as we might have expected, that the energy of interaction of polarization with the electric field is a consequence of orientation rather than position.

In order to find the expression for the field energy when the fields are *internal* to the system of charge and polarization we return to our general formula (5.1.3) for a system of point charges, which we write in the form

$$\begin{aligned}\mathcal{U} = {} & \tfrac{1}{2}q^{(1)}\left(\frac{q^{(2)}}{r^{12}} + \frac{q^{(3)}}{r^{13}} + \cdots + \frac{q^{(N)}}{r^{1N}}\right) \\ & + \tfrac{1}{2}q^{(2)}\left(\frac{q^{(1)}}{r^{21}} + \frac{q^{(3)}}{r^{23}} + \cdots + \frac{q^{(N)}}{r^{2N}}\right) \\ & + \tfrac{1}{2}q^{(3)}\left(\frac{q^{(1)}}{r^{31}} + \frac{q^{(2)}}{r^{32}} + \frac{q^{(4)}}{r^{34}} + \cdots + \frac{q^{(N)}}{r^{3N}}\right) + \cdots \\ & + \tfrac{1}{2}q^{(N)}\left(\frac{q^{(1)}}{r^{N1}} + \frac{q^{(2)}}{r^{N2}} + \frac{q^{(3)}}{r^{N3}} + \cdots + \frac{q^{(N-1)}}{r^{N(N-1)}}\right),\end{aligned} \tag{5.2.4}$$

and we now let $q^{(2)}$ and $q^{(3)}$ constitute a dipole. In other words we let $q^{(2)} = -q^{(3)}$, as shown in Figure 18, and proceed to the limit in the usual way, but with the provision that terms containing r^{23} are ignored as internal

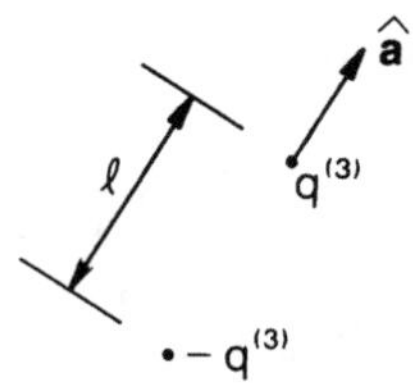

FIGURE 18. Formation of electric dipole moment from two point charges $q^{(2)}$ and $q^{(3)}$.

self energy terms of the dipole in the same way, and for the same reason, that all terms containing such quantities as r^{11}, r^{22}, r^{33} etc. were ignored in the finite point charge case. This can be looked upon as redefining $\mathscr{U}$ so as to exclude the singular self energy of the point dipole. However, in this writer's view it is more appropriate and, indeed, more realistic to note that the point dipoles and point charges, from which the dipoles are defined, are only models which are convenient for developing the theory, as well as other purposes. In other words the point charge is simply a model of a dense distribution of charge over a very small but finite region, in which the total charge is taken as being at the centroid of the distribution. From this point of view the so-called singular self energy terms, which have been and always are omitted in the treatment, can actually be shown to vanish because the charge density, although large, is bounded. Furthermore, in the definition of the dipole moment the point charges $q^{(n)}$ do not actually become infinite and the separation l does not actually vanish. Nevertheless, it is mathematically very convenient to employ the point charge and point dipole model and omit the singularities that arise from the simplified model, as we do, because the charges and dipoles then exist at discrete points.

Proceeding in accordance with the foregoing, from (5.2.4) we obtain

$$\begin{aligned}\mathscr{U} = {} & \tfrac{1}{2}q^{(1)}\left[q^{(3)}\left(-\frac{1}{r^{12}}+\frac{1}{r^{13}}\right)+\frac{q^{(4)}}{r^{14}}+\cdots+\frac{q^{(N)}}{r^{1N}}\right] \\ & +\tfrac{1}{2}q^{(3)}\left[q^{(1)}\left(-\frac{1}{r^{21}}+\frac{1}{r^{31}}\right)+q^{(4)}\left(-\frac{1}{r^{24}}+\frac{1}{r^{34}}\right)+\cdots\right. \\ & \left.+q^{(N)}\left(-\frac{1}{r^{2N}}+\frac{1}{r^{3N}}\right)\right] \\ & +\tfrac{1}{2}q^{(4)}\left[\frac{q^{(1)}}{r^{41}}+q^{(3)}\left(-\frac{1}{r^{42}}+\frac{1}{r^{43}}\right)+\cdots+\frac{q^{(N)}}{r^{1N}}\right]+\cdots \\ & +\tfrac{1}{2}q^{(N)}\left[\frac{q^{(1)}}{r^{N1}}+q^{(3)}\left(-\frac{1}{r^{N2}}+\frac{1}{r^{N3}}\right)+\cdots+\frac{q^{(N-1)}}{r^{N(N-1)}}\right],\end{aligned} \tag{5.2.5}$$

and since

$$\frac{1}{r^{n3}}=\frac{1}{r^{n2}}+l\tilde{\mathbf{a}}\cdot\nabla^2\left(\frac{1}{r^{n2}}\right),\quad \nabla^2\left(\frac{1}{r^{n2}}\right)=-\nabla^n\left(\frac{1}{r^{n2}}\right), \tag{5.2.6}$$

Eq. (5.2.5) can be written in the form

$$\mathcal{U} = \tfrac{1}{2}q^{(1)}\left[q^{(3)}l\hat{\mathbf{a}}\cdot\nabla^2\left(\frac{1}{r^{12}}\right) + \frac{q^{(4)}}{r^{14}} + \cdots + \frac{q^{(N)}}{r^{1N}}\right]$$
$$+ \tfrac{1}{2}q^{(3)}\left[q^{(1)}l\hat{\mathbf{a}}\cdot\nabla^2\left(\frac{1}{r^{21}}\right) + q^{(4)}l\hat{\mathbf{a}}\cdot\nabla^2\left(\frac{1}{r^{24}}\right) + \cdots + q^{(N)}l\hat{\mathbf{a}}\cdot\nabla^2\left(\frac{1}{r^{2N}}\right)\right] + \cdots$$
$$+ \tfrac{1}{2}q^{(N)}\left[\frac{q^{(1)}}{r^{N1}} + q^{(3)}l\hat{\mathbf{a}}\cdot\nabla^2\left(\frac{1}{r^{N2}}\right) + \cdots + \frac{q^{(N-1)}}{r^{N(N-1)}}\right], \tag{5.2.7}$$

which with $\mathbf{m}^{(2)} = q^{(3)}l\hat{\mathbf{a}}$ enables us to write

$$\mathcal{U} = \tfrac{1}{2}q^{(1)}\left[\mathbf{m}^{(2)}\cdot\nabla^2\left(\frac{1}{r^{12}}\right) + \frac{q^{(4)}}{r^{14}} + \cdots + \frac{q^{(N)}}{r^{1N}}\right]$$
$$+ \tfrac{1}{2}\mathbf{m}^{(2)}\cdot\nabla^2\left[\frac{q^{(1)}}{r^{21}} + \frac{q^{(4)}}{r^{24}} + \cdots + \frac{q^{(N)}}{r^{2N}}\right] + \cdots$$
$$+ \tfrac{1}{2}q^{(N)}\left[\frac{q^{(1)}}{r^{N1}} + \mathbf{m}^{(2)}\cdot\nabla^2\left(\frac{1}{r^{N2}}\right) + \cdots + \frac{q^{(N-1)}}{r^{N(N-1)}}\right]. \tag{5.2.8}$$

Now, if we introduce an additional dipole at, say, $q^{(5)}$, then $q^{(5)} = -q^{(6)}$ and

$$\frac{1}{r^{n6}} = \frac{1}{r^{n5}} + \lambda\hat{\boldsymbol{\alpha}}\cdot\nabla^5\left(\frac{1}{r^{n5}}\right), \quad \nabla^5\left(\frac{1}{r^{n5}}\right) = -\nabla^n\left(\frac{1}{r^{n5}}\right), \tag{5.2.9}$$

from which, with $\mathbf{m}^{(5)} = q^{(6)}\lambda\hat{\boldsymbol{\alpha}}$, it is clear that Eq. (5.2.8) may be written

$$\mathcal{U} = \tfrac{1}{2}q^{(1)}\left[\mathbf{m}^{(2)}\cdot\nabla^2\left(\frac{1}{r^{12}}\right) + \frac{q^{(4)}}{r^{14}} + \mathbf{m}^{(5)}\cdot\nabla^5\left(\frac{1}{r^{15}}\right) + \cdots + \frac{q^{(N)}}{r^{1N}}\right]$$
$$+ \tfrac{1}{2}\mathbf{m}^{(2)}\cdot\nabla^2\left[\frac{q^{(1)}}{r^{21}} + \frac{q^{(4)}}{r^{24}} + \mathbf{m}^{(5)}\cdot\nabla^5\left(\frac{1}{r^{25}}\right) + \cdots + \frac{q^{(N)}}{r^{2N}}\right]$$
$$+ \tfrac{1}{2}q^{(4)}\left[\frac{q^{(1)}}{r^{41}} + \mathbf{m}^{(2)}\cdot\nabla^2\left(\frac{1}{r^{42}}\right) + \mathbf{m}^{(5)}\cdot\nabla^5\left(\frac{1}{r^{45}}\right) + \cdots + \frac{q^{(N)}}{r^{4N}}\right]$$
$$+ \tfrac{1}{2}\mathbf{m}^{(5)}\cdot\nabla^5\left[\frac{q^{(1)}}{r^{51}} + \mathbf{m}^{(2)}\cdot\nabla^2\left(\frac{1}{r^{52}}\right) + \frac{q^{(4)}}{r^{54}} + \cdots + \frac{q^{(N)}}{r^{5N}}\right] + \cdots$$
$$+ \tfrac{1}{2}q^{(N)}\left[\frac{q^{(1)}}{r^{N1}} + \mathbf{m}^{(2)}\cdot\nabla^2\left(\frac{1}{r^{N2}}\right) + \frac{q^{(4)}}{r^{N4}} + \mathbf{m}^{(5)}\cdot\nabla^5\left(\frac{1}{r^{N5}}\right) + \cdots + \frac{q^{(N-1)}}{r^{N-1}}\right]. \tag{5.2.10}$$

An examination of this expression and the previous expressions (2.1.13), (2.1.14) and (3.1.6) for φ and $\mathbf{E}$ for systems of point charges and point dipoles reveals that for a system of N point charges and D point dipoles we may write

$$\mathcal{U} = \frac{1}{2}\sum_{n=1}^{N} q^{(n)}\varphi^{(n)} - \frac{1}{2}\sum_{d=1}^{D} \mathbf{m}^{(d)}\cdot\mathbf{E}^{(d)}. \tag{5.2.11}$$

When we proceed from the discrete point charge and dipole point of view to the distributed charge density and polarization density (or field) point of view, we must remember that the electric field **E** in the polarized region is not the force on a point charge because the local surface term

$$\mathbf{E}^L = -\int_\Sigma \mathbf{n}\cdot\mathbf{P}\nabla^P\left(\frac{1}{r_1}\right)dS, \tag{5.2.12}$$

which depends on the shape of the surface Σ surrounding the vanishingly small element of volume, has been excluded from the vector **E** appearing in the field equations. As a consequence we can expect the energy associated with $\mathbf{E}^L$ to be excluded when we pass to the distributed polarization density-field point of view and replace $\mathbf{E}^d$ by **E** in the expression for $\mathcal{U}$. But this energy associated with $\mathbf{E}^L$, which will be left out, is purely local, i.e., is electric energy of the matter in the differential element of volume.

In addition to the local energy associated with the Lorentz local field, which energy must be quadratic in the polarization, there is another very important contribution to the local energy, which we have not yet discussed. This energy is the energy of induced polarization and arises by virtue of the fact that, in general, at each point of the body, the electric field has generated the polarization. In other words, when there was no electric field at a point P, there was no polarization, and when the electric field **E** arose at P it moved microscopic charge around in the matter, slightly displacing positive and negative charge and thereby polarizing the matter in the neighborhood of P. In our development of the theory up to this point we have assumed the polarization **P** to exist and not considered the energy necessary to create **P**. Since, as we have already mentioned, **P** at P is created by **E** at P, we must add this energy of induced polarization in addition to the local energy associated with the Lorentz local field $\mathbf{E}^L$, when we proceed from the point charge and point dipole point of view to the distributed charge density and polarization density, or field, point of view. This combined local energy is directly analogous to the stored elastic energy of a mechanical system. Let us denote this combined local energy by u^L.

Now, let us determine the expression for the energy when we have a dielectric cavity in a metal, where the cavity has distributed charge density ρ and polarization **P**. This is the general case because putting the metal outside the dielectric produces surface charge σ on the metal-dielectric interface and at the same time eliminates consideration of any fields outside the dielectric, which gives us the simplest general problem we can consider. Thus, we consider the configuration shown in Figure 19. Since we have an electric field and polarization in the insulating cavity and no electric field or polarization in the surrounding metal, on account of Eq. (3.5.10) we must have charge density σ on the interface. Consequently, when we apply Eq. (5.2.11) for electrostatic field energy to the situation under consideration in Figure 19 and employ the argument to go from the discrete model to the

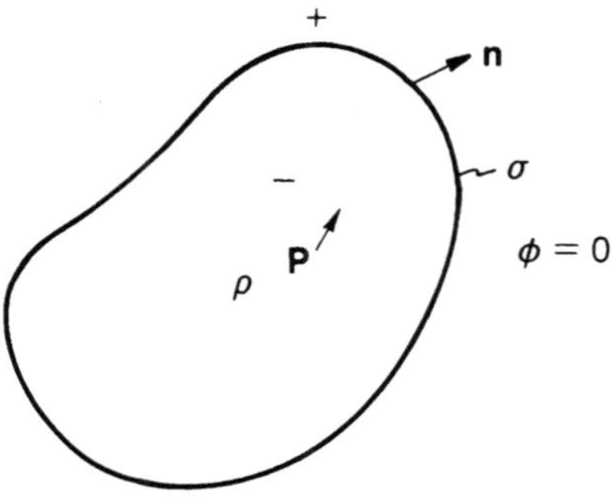

FIGURE 19. Interior dielectric in a metal.

continuous distribution we obtain

$$\mathscr{U} = \frac{1}{2}\int_S \sigma\varphi\, dS + \frac{1}{2}\int_V \rho\varphi\, dV - \frac{1}{2}\int_V \mathbf{P}\cdot\mathbf{E}\, dV + \int_V u^L(\mathbf{P})\, dV, \tag{5.2.13}$$

where u^L is included, in accordance with our discussion, in order to account for the local electrical energy of induced polarization and Lorentz local field, which is not included in **E**. The boundary conditions at the metal-vacuum interface are given by Eqs. (3.5.10) and (3.5.18), which for the situation under consideration here take the forms

$$\mathbf{n}\cdot(\mathbf{E}^+ - \mathbf{D}^-) = 4\pi\sigma, \tag{5.2.14}$$

$$\mathbf{n}\times(\mathbf{E}^+ - \mathbf{E}^-) = 0, \tag{5.2.15}$$

and since $\mathbf{E}^+ \equiv 0$ in the conductor, we find

$$\sigma = -\frac{\mathbf{n}\cdot\mathbf{D}^-}{4\pi}, \tag{5.2.16}$$

$$\mathbf{n}\times\mathbf{E}^-_{\cdot} = 0, \tag{5.2.17}$$

and from (3.2.16) we have

$$\mathbf{E}^- = -\nabla\varphi^-. \tag{5.2.18}$$

Substituting in the expression for $\mathscr{U}$ and dropping the $-$, we obtain

$$\mathscr{U} = -\frac{1}{8\pi}\int_S n_i D_i \varphi + \frac{1}{2}\int_V (\rho\varphi + P_i\varphi_{,i})\, dV + \int_V u^L(\mathbf{P})\, dV. \tag{5.2.19}$$

which, with the divergence theorem, yields

$$\mathscr{U} = \frac{1}{2}\int_V \left[-\frac{D_{i,i}}{4\pi}\varphi - \frac{D_i}{4\pi}\varphi_{,i} + \rho\varphi + P_i\varphi_{,i} \right] dV + \int_V u^L(\mathbf{P})\, dV. \tag{5.2.20}$$

This last expression with (3.3.7) may be written

$$\mathscr{U} = \frac{1}{2}\int_V \left[\left(-\frac{D_{i,i}}{4\pi} + \rho\right)\varphi - \left(\frac{E_i}{4\pi} + P_i\right)\varphi_{,i} + P_i\varphi_{,i}\right] dV + \int_V u^L(\mathbf{P})\,dV, \tag{5.2.21}$$

from which with (3.3.8) and (3.3.10) we obtain

$$\mathscr{U} = \frac{1}{8\pi}\int_V E_iE_i\,dV + \int_V u^L(\mathbf{P})\,dV. \tag{5.2.22}$$

This last expression in (5.2.22) shows that in the most general case, i.e., when ρ, $\mathbf{P}$ and σ are present, the stored energy in the field and the matter is just the ordinary electrostatic energy of free space $E^2/8\pi$ and the local material energy of induced polarization plus the energy associated with the local field that has been left out in defining $\mathbf{E}$.

5.3 Energy in Linear Electrostatics

The expression (5.2.22), which we have obtained, is perfectly general and is valid in the nonlinear case. However, it should be noted that in linear electrostatics we have

$$u^L = \frac{1}{2\chi} P_iP_i, \tag{5.3.1}$$

$$E_i = \frac{1}{\chi} P_i, \tag{5.3.2}$$

in the isotropic case and

$$u^L = \tfrac{1}{2}\beta_{ij}P_iP_j \tag{5.3.3}$$

$$E_i = \beta_{ij}P_j \tag{5.3.4}$$

in the anisotropic case, where β_{ij} is the reciprocal of χ_{ij}. In either case we have

$$u^L = \tfrac{1}{2}E_iP_i, \tag{5.3.5}$$

which with (5.2.22) enables us to write

$$\mathscr{U} = \frac{1}{2}\int_V \left[\frac{E_iE_i}{4\pi} + E_iP_i\right] dV = \frac{1}{8\pi}\int_V E_iD_i\,dV, \tag{5.3.6}$$

which is the well-known expression for the energy in *linear* electrostatics.

Note that as a consequence of the existence of the stored energy function $\mathscr{U}$ in linear electrostatics, the tensor β_{ij} and, of course, its reciprocal χ_{ij} are symmetric, i.e.,

$$\beta_{ij} = \beta_{ji}, \quad \chi_{ij} = \chi_{ji}, \tag{5.3.7}$$

since in the sum in (5.3.3) the terms occur in pairs such as, e.g.,

$$\tfrac{1}{2}(\beta_{12}+\beta_{21})E_1E_2,$$

which shows that only the symmetric combination counts, and, hence, we may take $\beta_{12}=\beta_{21}$ without loss in generality. This ends our development of electrostatics and brings us to magnetostatics.

II

MAGNETOSTATICS

CHAPTER 6

Magnetic Field Equations in Regions Carrying Steady Currents

6.1 The Magnetic Force Law

Before we introduce the magnetic force law we should make a number of observations. Many of the older texts develop magnetostatics by defining magnetic point charges and introducing Coulomb's law for the magnetic charges, and developing magnetostatics in direct analogy with electrostatics. This is obviously a very simple and straightforward approach. However, it is open to the following rather devastating criticisms:

1. Magnetic charges are known not to exist, i.e., only the + and − charge in combination as the magnetic dipole moment has ever been found no matter how small the element of matter examined.
2. Magnetism is known to be due to the motion of electric charge, i.e., the flow of current.

Consequently, in order for our basic laws to correspond, at least in some sense, to physical reality, we cannot develop magnetism from the point magnetic charge point of view, but must consider the interaction of moving electric charges (or current elements).

We start the treatment with Amperes law of force between the circuit elements shown in Figure 20 in the form

$$d\mathbf{F}^{(PQ)} = \frac{\gamma I^P\, d\mathbf{s}^P \times (I^Q\, d\mathbf{s}^Q \times \hat{\mathbf{r}}_1)}{r_1^2} = \frac{\gamma I^P I^Q}{r_1^2}\, d\mathbf{s}^P \times (d\mathbf{s}^Q \times \hat{\mathbf{r}}_1), \qquad (6.1.1)$$

where $d\mathbf{F}^{(PQ)}$ is the force exerted by the element $d\mathbf{s}^Q$ of the Q circuit on the element $d\mathbf{s}^P$ of the P circuit, γ is introduced for dimensional purposes, I^P denotes the electric charge per unit time or current flowing in the P circuit and I^Q the corresponding quantity flowing in the Q circuit. It should be noted that since the current in the circuit I represents the total electric charge flowing per unit time in the cicuit, we can write

$$I^Q\, d\mathbf{s}^Q = N^Q q^Q \mathbf{v}^Q, \quad I^P\, d\mathbf{s}^P = N^P q^P \mathbf{v}^P, \qquad (6.1.2)$$

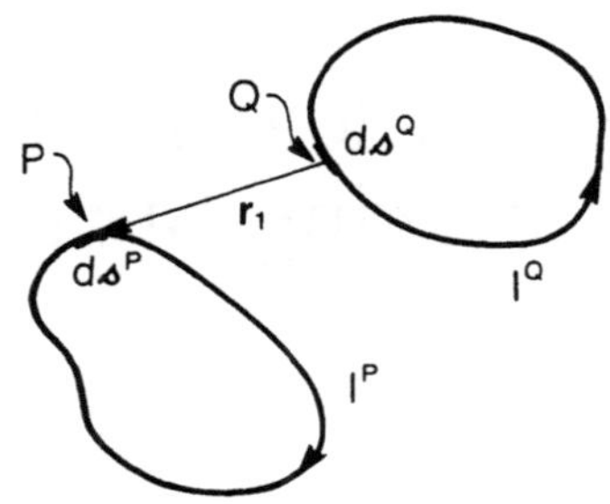

FIGURE 20. Force between interacting circuit elements.

where N represents the number of charges q moving with velocity $\mathbf{v}$. Since the circuits are hypothetical, it is enough for us to consider circuits which have one charge flowing in them, so that $N^P = N^Q = 1$. Moreover, since we are considering steady currents, I^Q is the same at every cross-section of the Q circuit and, hence, q^Q is moving with velocity $\mathbf{v}^Q$ at all points of the circuit simultaneously. Under those circumstances we can write the expression for $d\mathbf{F}^{(PQ)}$ in the form

$$d\mathbf{F}^{(PQ)} = \gamma \frac{q^P q^Q}{r_1^2} \mathbf{v}^P \times (\mathbf{v}^Q \times \hat{\mathbf{r}}_1). \tag{6.1.3}$$

It should be noted that neither of these equations, i.e. (6.1.1) or (6.1.3), were written in these forms by Ampere. They are, however, an analytical consequence of the results of Ampere's experiments and seem to me to be the best single statement for developing magnetostatics.

However, before proceeding with the development we must say something about the magnetic units in which we will work. Again, we proceed in the most straightforward way and set $\gamma = 1$, and then the form for $d\mathbf{F}$ in (6.1.1) gives us the electromagnetic unit of current, which clearly has the dimensions

$$I = T^{-1}(ML)^{1/2}. \tag{6.1.4}$$

However, the form for $d\mathbf{F}$ in (6.1.3) gives us the *electromagnetic* unit of electric charge and this differs from the *electrostatic* unit of electric charge both dimensionally and numerically. The e.m.u. of electric charge is a much larger unit than the e.s.u. From Coulombs law in (2.1.3) we already have

$$[q^e] = L^{3/2}M^{1/2}T^{-1} = \frac{L}{T}(ML)^{1/2}. \tag{6.1.5}$$

From (6.1.3) we find

$$[q^M] = \left[\frac{ML}{T^2}\frac{L^2}{L^2}T^2\right]^{1/2} = (ML)^{1/2}. \tag{6.1.6}$$

Hence, in units the ratio is

$$\frac{[q^e]}{[q^M]} = \frac{L}{T}, \tag{6.1.7}$$

which shows that the ratio is a velocity. Moreover, measurement shows that this velocity is the speed of light $c = 3 \times 10^{10}$ cm/sec. It should also be noted that the e.m.u. of current corresponds to 10 amps.

At this point it should be noted that the differential form of Ampere's law which we have given in (6.1.1) and (6.1.3) violates Newton's third law of action and reaction for each set of elements $d\mathbf{s}^Q$ and $d\mathbf{s}^P$, since successive applications of the forms given show that

$$d\mathbf{F}^{(PQ)} \neq -d\mathbf{F}^{(QP)}. \tag{6.1.8}$$

However, when the *total* force $\mathbf{F}^P$ is determined by integrating over *both* complete current loops, we find that the law of action and reaction is satisfied for the interaction of the two complete circuits. This can readily be seen by using the vector identity

$$\mathbf{a} \times (\mathbf{b} \times \mathbf{c}) = (\mathbf{a}\cdot\mathbf{c})\mathbf{b} - (\mathbf{a}\cdot\mathbf{b})\mathbf{c},$$

and substituting in (6.1.1) and integrating over both current loops to obtain

$$\mathbf{F}^{(PQ)} = I^P I^Q \oint_{C^P} \oint_{C^Q} \left[\frac{d\mathbf{s}^P \cdot \mathbf{r}_1}{r_1^3} d\mathbf{s}^Q - \frac{d\mathbf{s}^P \cdot d\mathbf{s}^Q \mathbf{r}_1}{r_1^3} \right]. \tag{6.1.9}$$

Now, since

$$\nabla^P \left(\frac{1}{r_1} \right) = -\frac{\mathbf{r}_1}{r_1^3}, \tag{6.1.10}$$

we see that the first integral can be written

$$-\oint_{C^Q} d\mathbf{s}^Q \oint_{C^P} d\mathbf{s}^P \cdot \nabla^P \left(\frac{1}{r_1} \right) = -\oint_{C^Q} d\mathbf{s}^Q \oint_{C^P} d\left(\frac{1}{r_1} \right) = -\oint_{C^Q} d\mathbf{s}^Q \left[\frac{1}{r_1} \right]_0^0 = 0. \tag{6.1.11}$$

Hence the expression for the resultant force exerted on circuit P by circuit Q may be written

$$\mathbf{F}^{(PQ)} = -I^P I^Q \oint_{C^P} \oint_{C^Q} \frac{d\mathbf{s}^P \cdot d\mathbf{s}^Q \mathbf{r}_1}{r_1^3}, \tag{6.1.12}$$

from which it is clear that

$$\mathbf{F}^{(PQ)} = -\mathbf{F}^{(QP)}, \tag{6.1.13}$$

which says that the resultant force acting between the two complete circuits is equal and opposite.

Although we have shown that the resultant force arising from the first term in brackets on the right-hand side of (6.1.9) due to any given element $d\mathbf{s}^Q$ of the Q circuit acting on the entire P circuit vanishes by virtue of (6.1.11), the forces due to the individual elements of circuit P can still result in a couple. However, since the resultant force vanishes, the moment of the force system due to an element of the Q circuit on the entire P circuit about any

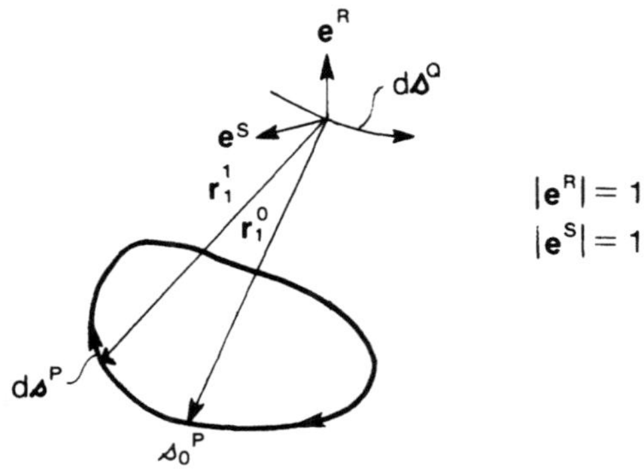

FIGURE 21. Schematic diagram showing element of Q circuit and entire P circuit.

point is the same. Hence, we can take moments about the most convenient point we can find, which from the integrand in (6.1.11) and the diagram in Figure 21 is the point of the fixed element $d\mathbf{s}^Q$ of the source. Accordingly, we write

$$\begin{aligned} d\mathbf{C} &= -\oint_{C^P} (-\mathbf{r}_1) \times \left(d\mathbf{s}^Q d\mathbf{s}^P \cdot \nabla^P \left(\frac{1}{r_1} \right) \right) \\ &= \oint_{C^P} \mathbf{r}_1 \times d\left(\frac{1}{r_1} \right) d\mathbf{s}^Q = -\oint_{C^P} \frac{dr_1}{r_1} \hat{\mathbf{r}}_1 \times d\mathbf{s}^Q \\ &= -\oint_{C^P} \frac{dr_1}{r_1} \hat{\mathbf{r}}_1 \times \mathbf{e}^Q \, ds^Q, \end{aligned} \tag{6.1.14}$$

where $\mathbf{e}^Q$ is a fixed unit vector in the direction of $d\mathbf{s}^Q$. Since $\hat{\mathbf{r}}_1$ is a unit vector that varies as we traverse the P circuit while $\mathbf{e}^Q$ is a fixed unit vector, we may define fixed orthogonal unit vectors $\mathbf{e}^R$ and $\mathbf{e}^S$ lying in the plane normal to $\mathbf{e}^Q$, which enables us to write (6.1.14) in the form

$$d\mathbf{C} = -\oint_{C^P} \frac{ds}{r_1} [g^R(s)\mathbf{e}^R + g^S(s)\mathbf{e}^S] \, ds^Q, \tag{6.1.15}$$

in which we have defined the s coordinate as the distance from s_0^P along the curve of the P circuit and we have replaced dr_1 by ds since they are identical. Moreover, since $r_1 = r_1(s)$ we can write (6.1.15) in the form

$$\begin{aligned} d\mathbf{C} &= -\oint_{C^P} [f^R(s)\mathbf{e}^R + f^S(s)\mathbf{e}^S] \, ds \, ds^Q \\ &= [F^R(s)\mathbf{e}^R + F^S(s)\mathbf{e}^S]_0^0 \, ds^Q = 0, \end{aligned} \tag{6.1.16}$$

since F^R and F^S are single-valued. Thus we have shown that the couple arising from the force term in (6.1.11) due to any given element $d\mathbf{s}^Q$ of the Q circuit acting on the entire P circuit vanishes as well as the force.

In view of (6.1.16), Eq. (6.1.12) shows that the law of action and reaction is satisfied for the two complete circuits. Furthermore, the form of the integrand in (6.1.12) reveals that for that expression the law of action and reaction is satisfied differential element by differential element. As a matter of fact it should be noted that the present integral form of the law in (6.1.12) is much closer to the results of Ampere's experiments than the previous differential form in (6.1.1). However, the differential form is required to obtain the field theory. In particular we are ultimately interested in the form of the law in (6.1.3), i.e.,

$$d\mathbf{F}^{(PQ)} = q^P \mathbf{v}^P \times \frac{(q^Q \mathbf{v}^Q \times \mathbf{r}_1)}{r_1^3}, \tag{6.1.17}$$

and eventually when the currents are not steady in closed loops. The fact that this form does not satisfy the law of action and reaction by itself does not deter us because, as we have seen, in the case of steady currents, which must flow in closed loops, the *resultant* forces between entire circuits does satisfy the law of action and reaction and in the nonsteady case we have time varying magnetic (and electric fields) and as we shall see later in electromagnetism the electromagnetic field carries linear momentum (even in free space).

6.2 The Magnetic Induction Field

We now define the magnetic induction field **B** at P due to current $q^Q \mathbf{v}^Q$ at Q by the equation

$$\mathbf{B}(P) = \frac{q^Q \mathbf{v}^Q \times \mathbf{r}_1}{r_1^3}, \tag{6.2.1}$$

from which it is clear that **B** is not a polar but an axial vector. In other words the direction of **B** depends on whether the coordinate system is left or right handed. We are assuming a right-handed coordinate system throughout. In the equation of definition we have assumed that we have a point charge q^Q moving with a velocity $\mathbf{v}^Q$. For convenience in the sequel let

$$\mathbf{w}^Q = q^Q \mathbf{v}^Q, \quad \mathbf{w}^P = q^P \mathbf{v}^P, \tag{6.2.2}$$

then

$$\mathbf{B} = \frac{\mathbf{w}^Q \times \mathbf{r}_1}{r_1^3}, \tag{6.2.3}$$

and from (6.1.17), we have

$$d\mathbf{F}^P = \mathbf{w}^P \times \mathbf{B}, \tag{6.2.4}$$

where, in terms of the circuit loop variables, from (6.1.2) we have

$$\mathbf{w}^Q = I^Q d\mathbf{s}^Q, \quad \mathbf{w}^P = I^P d\mathbf{s}^P. \tag{6.2.5}$$

Now, it is clear from (6.2.1) and (6.1.10) that we may write

$$\mathbf{B} = -\mathbf{w}^Q \times \nabla^P\left(\frac{1}{r_1}\right) = \nabla^P\left(\frac{1}{r_1}\right) \times \mathbf{w}^Q, \tag{6.2.6}$$

or

$$\mathbf{B} = \nabla^P \times \frac{\mathbf{w}^Q}{r_1} = \nabla^P \times \mathbf{A}, \tag{6.2.7}$$

where

$$\mathbf{A} = \mathbf{w}^Q/r_1, \tag{6.2.8}$$

is the magnetic vector potential of the current element $\mathbf{w}^Q$ at Q. Clearly then from (6.2.7), (6.2.8) and (2.1.11), we have

$$\nabla^P \cdot \mathbf{B} = 0, \quad \text{except at } Q, \tag{6.2.9}$$

$$(\nabla^P)^2 \mathbf{A} = 0, \quad \text{except at } Q. \tag{6.2.10}$$

Although we cannot define the energy of a *magnetostatic* system by considering moving elements of charge in which the law of action and reaction is not satisfied, we can define the energy for a *magnetostatic* system by considering *steady* currents flowing in closed loops. This is not only not restrictive but is as expected because, as already mentioned, in the non-time dependent case all current elements must ultimately flow in closed loops. Thus, to define the energy of a magnetostatic system, we return to the expression (6.1.12) for the total force $\mathbf{F}^{(PQ)}$ between two closed loops, which with (6.1.10) may be written in the form

$$\mathbf{F}^{(PQ)} = I^P I^Q \oint_{C^P} \oint_{C^Q} d\mathbf{s}^P \cdot d\mathbf{s}^Q \nabla^P\left(\frac{1}{r_1}\right). \tag{6.2.11}$$

The form of this last equation is very suggestive. However, we cannot proceed in the simple way we did in electrostatics because, for one thing, the circuit loops have a finite extent and the point charges do not. Consequently, we consider the scalar

$$U^1 = I^P I^Q \oint_{C^P} \oint_{C^Q} \frac{d\mathbf{s}^P \cdot d\mathbf{s}^Q}{r_1}, \tag{6.2.12}$$

and displace the entire P circuit relative to the Q circuit through a pure infinitesimal rigid translation without rotation, and from the new scalar

$$U^2 = I^P I^Q \oint_{C^P} \oint_{C^Q} \frac{d\mathbf{s}^P \cdot d\mathbf{s}^Q}{r_2}, \tag{6.2.13}$$

where for each point P in the P circuit we have

$$\frac{1}{r_2} = \frac{1}{r_1} + d\mathbf{r}^Q \cdot \nabla^P\left(\frac{1}{r_1}\right), \tag{6.2.14}$$

where $d\mathbf{r}^{P}$ represents *the arbitrary* infinitesimal rigid translation of the P circuit relative to the Q circuit. Then

$$U^2 - U^1 = dU = d\mathbf{r}^{P}\cdot\nabla^P U = d\mathbf{r}^{P}\cdot\left[I^P I^Q \oint_{C^P}\oint_{C^Q} d\mathbf{s}^P\cdot d\mathbf{s}^Q \nabla^P\left(\frac{1}{r_1}\right)\right], \tag{6.2.15}$$

from which with (6.2.11) we obtain

$$\nabla^P U = I^P I^Q \oint_{C^P}\oint_{C^Q} d\mathbf{s}^P\cdot d\mathbf{s}^Q \nabla^P\left(\frac{1}{r_1}\right) = \mathbf{F}^{(PQ)}, \tag{6.2.16}$$

where $\mathbf{F}^{(PQ)}$ is the resultant force exerted by current loop Q on current loop P. The quantity $U \equiv U^1$ we identify the magnetostatic energy since its *translational* gradient with respect to the P points yields the force $\mathbf{F}^{(PQ)}$, which is exerted by current loop Q on current loop P. Then with the convention in (6.2.5) we have

$$U = \oint_{C^P}\oint_{C^Q} \frac{\mathbf{w}^P\cdot\mathbf{w}^Q}{r_1}, \tag{6.2.17}$$

which, from the circulation integral of (6.2.8) around the Q circuit, may be written in the form

$$U = \oint_{C^P} \mathbf{w}^P\cdot\mathbf{A}, \tag{6.2.18}$$

where

$$\mathbf{A} = \oint_{C^Q} \frac{\mathbf{w}^Q}{r_1}. \tag{6.2.19}$$

If we regard $\mathbf{w}^P$ as an element of a current loop, without completing the loop integral, we may write

$$u^M = \mathbf{w}^P\cdot\mathbf{A}, \tag{6.2.20}$$

where $\mathbf{A}$ in (6.2.20) may be due to the entire Q circuit or any portion thereof. The expression in (6.2.20) is directly analogous to the expression

$$u^E = q^P\varphi, \tag{6.2.21}$$

for the electrostatic energy of a point charge at P in a potential φ resulting from other point charges at Q points or rather charge density.

Let us now consider a system of current loops all distinct from each other and P as shown in Figure 22, and by superposition determine the magnetic induction vector $\mathbf{B}$ and vector potential $\mathbf{A}$ at P due to the elements of current of the loops at points Q^N. Thus

$$\mathbf{B}(P) = \sum_{n=1}^{N} \nabla^P\left(\frac{1}{r_1^{(n)}}\right) \times \mathbf{w}^{(n)}, \tag{6.2.22}$$

$$\mathbf{A}(P) = \sum_{n=1}^{N} \frac{\mathbf{w}^{(n)}}{r_1^{(n)}}, \tag{6.2.23}$$

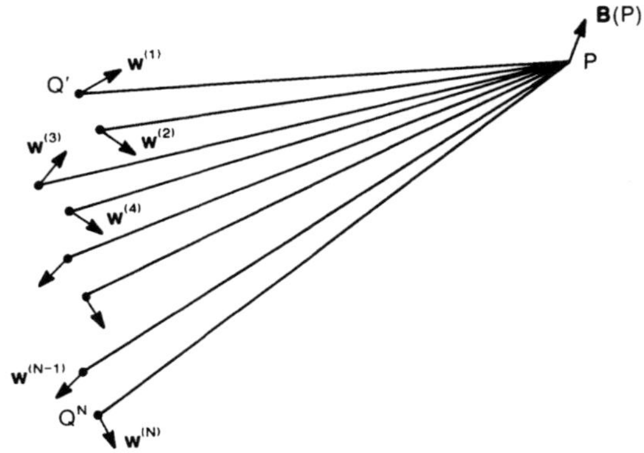

FIGURE 22. System of elements of current loops.

and

$$\mathbf{B}(P) = \nabla^P \times \sum_{n=1}^{N} \frac{\mathbf{w}^{(n)}}{r_1^{(n)}} = \nabla^P \times \mathbf{A}, \tag{6.2.24}$$

at each point P not occupied by current elements. Clearly, we have

$$\nabla^2\mathbf{A} = 0, \quad \nabla\cdot\mathbf{B} = 0 \quad \text{except at each } Q^{(n)}. \tag{6.2.25}$$

6.3 Continuous Distributions of Current

Before proceeding with a discussion of continuous distributions of current it should be noted that when we write the expression for **A** and **B** as sums over current elements $\mathbf{w}^{(n)}$, rather than sums over current loop integrals around $C^{(n)}$'s, we have either implicitly generalized the theory or are at liberty to explicitly do so. In other words we have effectively assumed that the expression for force exerted by an element of one current loop on an element of another current loop, also holds for the force exerted by an element of a current loop on another element of the same current loop. This is a logical physical generalization since it simply says that a given current element cannot distinguish whether any portion of the **A** or **B** field it experiences is generated by a current element of the same loop or a different loop. This means that in the two loop case we started with current loops as shown in Figure 23, for which we write the expression for the force **F**(P) at element P of the P current loop in the form

$$\mathbf{F}(P) = \mathbf{w}^P \times \mathbf{B}(P), \tag{6.3.1}$$

where we have employed the definition in (6.2.5) and

$$\mathbf{B}(P) = \oint_{C^{\mathcal{Q}}} \frac{\mathbf{w}^{\mathcal{Q}} \times \mathbf{r}_1}{r_1^3}. \tag{6.3.2}$$

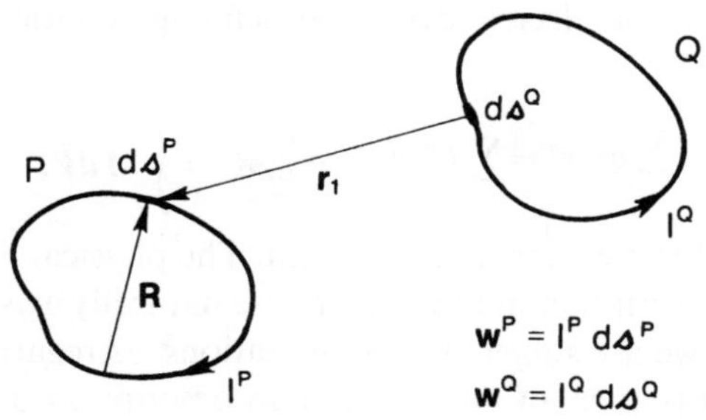

FIGURE 23. Two interacting current loops.

When we replaced the loop integrals by sums, we implicitly made the assumption that

$$\mathbf{B}(P) = \oint_{C^{\mathcal{Q}}} \frac{\mathbf{w}^{\mathcal{Q}} \times \mathbf{r}_1}{r_1^3} + \oint_{\substack{C^{\mathcal{P}} \\ R \neq 0}} \frac{\mathbf{w}^P \times \mathbf{R}}{R^3}, \tag{6.3.3)}$$

which was then written as

$$\mathbf{B}(P) = \sum_n \frac{\mathbf{w}^{(n)} \times \mathbf{r}_1^{(n)}}{(r_1^{(n)})^3}, \tag{6.3.4}$$

and the sum over n was meant to include both loops. Under these circumstances then we have

$$\mathbf{F}(P) = \mathbf{w}(P) \times \mathbf{B}(P), \tag{6.3.5}$$

where

$$\mathbf{B}(P) = \nabla^P \times \mathbf{A}(P), \tag{6.3.6}$$

and

$$\mathbf{A}(P) = \sum_n \frac{\mathbf{w}^{(n)}}{r_1^n}, \tag{6.3.7}$$

and the sum over n is to be interpreted as including current elements of the same loop as well as elements of other loops. As a consequence of this generalization it is clear that we will have to generalize our expression for magnetic energy in (6.2.18) along similar lines so as to account for what might well be called the finite portion of the magnetic self energy of a loop. This will be discussed in greater detail later on when we discuss the magnetic field energy of a collection of loops.

We now proceed to replace the discrete current elements $\mathbf{w}^{(n)}$, which must, of course, be portions of loops, by a variable distribution of current density $\mathbf{J}$, which may have discontinuities at points, lines and surfaces. This replacement takes place in exactly the same manner that the $q^{(n)}$ were replaced by ρ in Sec. 2.2 in the electric case. The defining equation for the current density is

$$\sum_n q^{(n)}\mathbf{v}^n = \sum_n I^{(n)}\, d\mathbf{s}^{(n)} = \sum_n \mathbf{w}^n = \int_V \mathbf{J}\, dV, \tag{6.3.8}$$

where V is the volume enclosing the current. The physical limitations on the distributed field description in the electric case naturally exist in the magnetic case also. Hence, we are subject to the limitations, as regards rate of change of variable with distance, of any continuum description of matter utilizing macroscopic laws. Then from (6.3.7) and (6.3.6), respectively, we may write the expressions

$$\mathbf{A}(P) = \int_V \frac{\mathbf{J}\, dV}{r_1}, \tag{6.3.9}$$

$$\mathbf{B}(P) = \nabla^P \times \mathbf{A} = \int_V \nabla^P\left(\frac{1}{r_1}\right) \times \mathbf{J}\, dV, \tag{6.3.10}$$

where P must be a point outside the region containing current density $\mathbf{J}$. Clearly then, we also have the relations

$$\nabla^P \cdot \nabla^P \mathbf{A} = 0, \quad \nabla^P \cdot \mathbf{B} = 0 \quad \text{except where } \mathbf{J} \neq 0. \tag{6.3.11}$$

We have now discussed the magnetic field and magnetic vector potential at points outside the region containing current density for a volume distribution of current density. We could now introduce a surface distribution of current density $\mathbf{K}$ in exactly the same way that we introduced a surface distribution of charge density σ. However, since we consider this to be unphysical, we will not bother to do it. Now, since the expressions for the magnetic potential and magnetic induction have the same dependence on r_1 as the analogous expressions for the electric potential and electric field, i.e., (2.2.2) and (2.2.3), and we have shown that we may use the same expressions for the electric potential and electric field at points inside the charged region, i.e., (2.4.4) and (2.4.9), because the improper integrals are convergent, it is clear that we may use the present expressions (6.3.9) and (6.3.10), respectively, for the magnetic vector potential and magnetic induction field *inside* the current region if $\mathbf{J}$ is bounded and integrable. Thus, we have

$$\mathbf{A}(P) = \int_V \frac{\mathbf{J}}{r_1}\, dV, \tag{6.3.12}$$

$$\mathbf{B}(P) = \int_V \nabla^P\left(\frac{1}{r_1}\right) \times \mathbf{J}\, dV = \nabla^P \times \int_V \frac{\mathbf{J}}{r_1}\, dV = \nabla^P \times \mathbf{A}(P), \tag{6.3.13}$$

at points inside the current region as well as outside. It is, of course, also clear that $\mathbf{A}$ and $\mathbf{B}$ are continuous.

Up to this point $\mathbf{J}$ has been bounded and integrable. During the remainder of the development we will need to use the divergence theorem, Stokes theorem and some other integral theorems analogous to the divergence theorem for some of our magnetic vector fields. Consequently, we will now show that the second spatial derivatives of **A** exist and are continuous at interior points if **J** and its first spatial derivatives are continuous. The proof follows the electric case given near the end of Sec. 2.4. Consider the expression

$$\nabla^P \mathbf{A} = \int_V \nabla^P\left(\frac{1}{r_1}\right)\mathbf{J}\,dV, \tag{6.3.14}$$

which, with (2.2.6), may be written

$$\nabla^P \mathbf{A} = -\int_V \nabla^Q\left(\frac{1}{r_1}\right)\mathbf{J}\,dV. \tag{6.3.15}$$

Equation (6.3.15) enables us to write

$$\nabla^P \mathbf{A} = -\int_V \left[\nabla^Q\left(\frac{\mathbf{J}}{r_1}\right) - \frac{\nabla^Q \mathbf{J}}{r_1}\right] dV. \tag{6.3.16}$$

We now make use of the integral theorem

$$\int_S \mathbf{n}\mathbf{G}\,dS = \int_V \nabla\mathbf{G}\,dV, \tag{6.3.17}$$

on the vector field G, which is a consequence of the divergence theorem,[17] which enables us to write (6.3.16) in the form

$$\nabla^P \mathbf{A} = -\int_S \mathbf{n}\frac{\mathbf{J}}{r_1}\,dS + \int_V \frac{\nabla^Q \mathbf{J}}{r_1}\,dV. \tag{6.3.18}$$

Since S is the surface bounding V and the point P is confined to the interior of V, the value of r_1 in the first integral does not vanish, and, hence, that integral has continuous derivatives of all orders. The second integral may be written

$$\int_V \frac{1}{r_1}\mathbf{e}_j \frac{\partial J_k}{\partial \xi_j}\mathbf{e}_k\,dV, \tag{6.3.19}$$

and the coefficient of each set of base vectors is a combination of integrals of the same form as

$$\int_V \frac{J_k}{r_1}\,dV, \tag{6.3.20}$$

which we know has continuous first derivatives. Hence, the expression for $\nabla^P \mathbf{A}$ has continuous first derivatives. Consequently, the divergence theorem

$$\int_S \mathbf{n}\cdot\nabla\mathbf{A}\,dS = \int_V \nabla\cdot\nabla\mathbf{A}\,dV, \tag{6.3.21}$$

is valid for S being *any* closed surface enclosing a volume V, within which **J** and ∇**J** are continuous.

At this point it should be noted that since in magnetostatics we are considering *steady* currents only, there can be no time dependence of any variable, and, in particular, there can be no time rate of increase of charge density at any macroscopic point. In other words any current flowing into an arbitrary volume must flow out. As a consequence, for an arbitrary volume element, we have

$$\int_S \mathbf{n}\cdot\mathbf{J}\,dS = 0, \tag{6.3.22}$$

and by the divergence theorem we find

$$\nabla\cdot\mathbf{J} = 0, \tag{6.3.23}$$

in magnetostatics.

6.4 Field Equation on Current Density

Consider the entire region carrying current to be enclosed within S_0 shown in Figure 24 and consider any surface S within which the current and its gradient are continuous. Integrate the normal component of $\nabla^P\mathbf{A}$ to obtain

$$\int_S \mathbf{n}\cdot\nabla^P\mathbf{A}\,dS, \tag{6.4.1}$$

which with (6.3.14) may be written

$$\int_S \mathbf{n}\cdot\nabla^P\mathbf{A}\,dS = \int_S \mathbf{n}\cdot\left[\int_{V_0}\nabla^P\left(\frac{1}{r_1}\right)\mathbf{J}\,dV\right]dS = -\int_S \mathbf{n}\cdot\left[\int_{V_0}\frac{\mathbf{r}_1}{r_1^3}\mathbf{J}\,dV\right]dS. \tag{6.4.2}$$

Interchanging the order of integration, we obtain

$$\int_S \mathbf{n}\cdot\nabla^P\mathbf{A}\,dS = -\int_{V_0}\int_S \frac{\mathbf{n}\cdot\mathbf{r}_1}{r_1^3}\mathbf{J}\,dS\,dV. \tag{6.4.3}$$

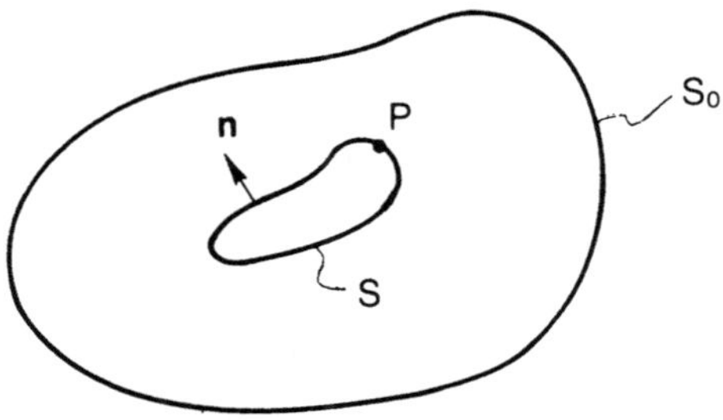

FIGURE 24. Entire region carrying current.

Integrating the right-hand side of (6.4.3) over S and employing (2.3.7)–(2.3.9), we obtain

$$\int_S \mathbf{n}\cdot\nabla^P \mathbf{A}\, dS = -4\pi \int_V \mathbf{J}\, dV, \tag{6.4.4}$$

since as shown in Sec. 2.3, for a closed surface S, the solid angle for any point outside S vanishes and for any point inside is 4π, and all of $V_0 - V$ is outside S. Since the integration variables are dummy variables and we restrict $\mathbf{J}$ to satisfy the conditions of the divergence theorem, we obtain

$$\int_V (\nabla^2 \mathbf{A} + 4\pi \mathbf{J})\, dV = 0, \tag{6.4.5}$$

for *any* volume V. Hence, we have

$$\nabla^2 \mathbf{A} = -4\pi \mathbf{J}, \tag{6.4.6}$$

at all points of the region carrying current. Equation (6.4.6), which is a vector Poisson equation, is the differential equation of magnetostatics in the absence of magnetization.

In a magnetic material, which may be a conductor such as iron or an insulator such as a ferrite, there can exist magnetic dipole moments, which are, microscopically speaking, a consequence of the spin and orbital angular momentum (or motion) of the electrons. In either case the magnetic dipole moments are due to the closed loop motion of electric charge. In ferromagnetic or ferrimagnetic materials the macroscopic magnetic dipole moments exist automatically, while in paramagnetic or diamagnetic materials the macroscopic magnetic dipole moments are induced by the macroscopic magnetic induction vector $\mathbf{B}$. Consequently, we determine the influence on the field vectors $\mathbf{A}$ and $\mathbf{B}$ of a magnetic dipole moment or magnetization in the next chapter.

CHAPTER 7

Magnetic Field Equations in Magnetized Regions Carrying Steady Current

7.1 The Magnetic Dipole Moment

We proceed in essentially the same way as in the electric case and first formally define a single magnetic dipole moment analytically from our present magnetic apparatus. To this end we consider a closed current loop in which a steady electric current I flows as shown in Figure 25. In all of our closed loop expressions for **B** and **A**, the loop integral

$$\oint_C f(r_1) I \, d\mathbf{s} = I \oint_C f(r_1) \, d\mathbf{s}, \tag{7.1.1}$$

exists. The application of a well-known integral theorem[18] yields

$$I \oint_C f \, d\mathbf{s} = I \int_S \mathbf{n} \times \nabla f \, d\mathbf{S}. \tag{7.1.2}$$

If the current loop is confined to a plane, **n** is constant and we have

$$I \oint_C f \, d\mathbf{s} = I\mathbf{n} \times \int_S \nabla f \, dS. \tag{7.1.3}$$

If, moreover, ∇f is constant we have

$$I \oint_C f \, d\mathbf{s} = I\mathbf{n}S_0 \times \nabla f, \tag{7.1.4}$$

where S_0 is the planar area within the loop. The quantity $IS_0\mathbf{n}$ is called the magnetic dipole moment of the steady planar current loop. We are particularly interested in the special case in which $I \to \infty$ and $S_0 \to 0$ while **n** maintains the fixed direction normal to the plane and the product IS_0 remains finite and equal to m. Thus we have

$$\mathbf{m} = \lim_{\substack{S_0 \to 0 \\ I \to \infty \\ \mathbf{n}\,\text{fixed}}} IS_0\mathbf{n} = m\mathbf{n}, \tag{7.1.5}$$

for the point magnetic dipole moment.

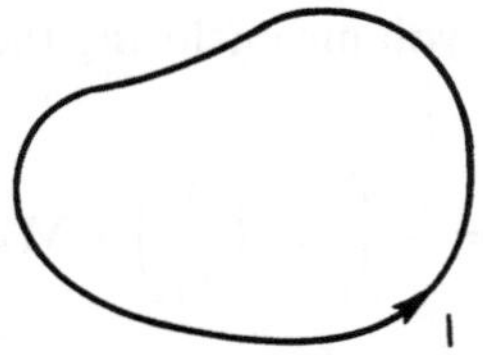

FIGURE 25. Closed current loop for definition of magnetic dipole moment.

We now wish to determine the magnetic vector potential **A** at P due to a dipole moment **m** at Q. To this end we consider the expression (6.2.19) with (6.2.5) for **A** at a point P due to a single closed current loop Q. Clearly

$$\mathbf{A} = I \oint_C \frac{d\mathbf{s}^Q}{r_1}, \tag{7.1.6}$$

and application of the integral theorem (7.1.2) yields

$$\mathbf{A} = I \oint_{S_0} \mathbf{n} \times \nabla^Q\left(\frac{1}{r_1}\right) dS = I\mathbf{n} \times \int_{S_0} \nabla^Q\left(\frac{1}{r_1}\right) dS, \tag{7.1.7}$$

and we take the limit as $S_0 \to 0$ and $I \to \infty$. Under these circumstances by the mean value theorem or Taylor's theorem, we can take $\nabla^Q(1/r_1)$ out of the integral and write

$$\mathbf{A} = \lim_{\substack{S_0 \to 0 \\ I \to \infty \\ \mathbf{n}\ \text{fixed}}} I\mathbf{n}S_0 \times \nabla^Q\left(\frac{1}{r_1}\right) = \mathbf{m} \times \nabla^Q\left(\frac{1}{r_1}\right)$$

$$= -\mathbf{m} \times \nabla^P\left(\frac{1}{r_1}\right) = \nabla^P\left(\frac{1}{r_1}\right) \times \mathbf{m}. \tag{7.1.8}$$

This expression for the magnetic vector potential **A** is for a single point magnetic dipole moment **m** at Q. Clearly, we may repeat this for a number of discrete point magnetic dipole moments at distinct points $Q^{(n)}$ just as we did in all the other discrete point cases in Secs. 2.1, 3.1 and 6.2 and by superposition obtain

$$\mathbf{A}(P) = \sum_{n=1}^{N} \mathbf{m}^{(n)} \times \nabla^Q\left(\frac{1}{r_1^{(n)}}\right). \tag{7.1.9}$$

We now proceed to replace the discrete magnetic dipole moments $\mathbf{m}^{(n)}$ by a variable distribution of magnetic dipole moment density **M**. This replacement takes place in exactly the same way that the electric dipole moments $\mathbf{m}^{(n)}$ were replaced by **P** in Sec. 3.1 in the electric case, and, of course, with the same limitations concerning the rate of variation with distance. The defining equation for the magnetization is

$$\sum_n \mathbf{m}^{(n)} = \int_V \mathbf{M}\, dV, \tag{7.1.10}$$

where V is the macroscopic volume enclosing the magnetization. Then from (7.1.9) we may write

$$\mathbf{A} = \int_V \mathbf{M} \times \nabla^Q\left(\frac{1}{r_1}\right) dV = -\int_V \nabla^Q\left(\frac{1}{r_1}\right) \times \mathbf{M}\, dV = \int_V \nabla^P\left(\frac{1}{r_1}\right) \times \mathbf{M}\, dV, \tag{7.1.11}$$

where the field point P must be outside the region containing the magnetization. Clearly, when the field point P is outside the magnetized region, r_1 never vanishes, and from (6.3.6) we have

$$\mathbf{B}(P) = \nabla^P \times \mathbf{A}^P = \int_V \nabla^P \times \left[\nabla^P\left(\frac{1}{r_1}\right) \times \mathbf{M}\right] dV. \tag{7.1.12}$$

7.2 Fields Inside the Magnetized Region

We now wish to determine expressions for the magnetic induction field **B** and the magnetic potential **A** in the interior of the magnetized region. In the case of the potential **A** it is perfectly clear, from the similarity in the dependence on r_1 of **A** due to a distribution of **M** in (7.1.11) and φ due to a distribution of P in (3.1.7), that we may use the present expression for **A** in (7.1.11) *within* the magnetized region. It is also perfectly clear, from the similarity in the dependence on r_1 of **B** due to a distribution of **M** in (7.1.12) and **E** due to a distribution of **P** in (3.1.9), that the present expression for **B** will *not* be valid *within* the magnetized region. However, it is quite clear that we may proceed in essentially the same way we did in redefining **E** in Section 3.2 and redefine the magnetic induction vector **B** so as to exclude the effect of the local field $\mathbf{B}^L$.

A convenient starting point is the expression for **A** in the form

$$\mathbf{A} = -\int_V \nabla^Q\left(\frac{1}{r_1}\right) \times \mathbf{M}\, dV, \tag{7.2.1}$$

which may be written

$$A = -\int_V \nabla^Q \times \left(\frac{\mathbf{M}}{r_1}\right) dV + \int_V \frac{\nabla^Q \times \mathbf{M}}{r_1} dV. \tag{7.2.2}$$

We now consider the region shown in Figure 26 in which we surround the field point P by a surface Σ enclosing a small volume V'', which we exclude from the domain of integration in the expression for **A** that we use in the determination of **B** and take the limit as $V'' \to 0$, as was done in Sec. 2.4 and 3.2. In other words we have

$$\mathbf{A}' = -\int_{V'} \nabla^Q \times \left(\frac{\mathbf{M}}{r_1}\right) dV + \int_{V'} \frac{\nabla^Q \times \mathbf{M}}{r_1} dV, \tag{7.2.3}$$

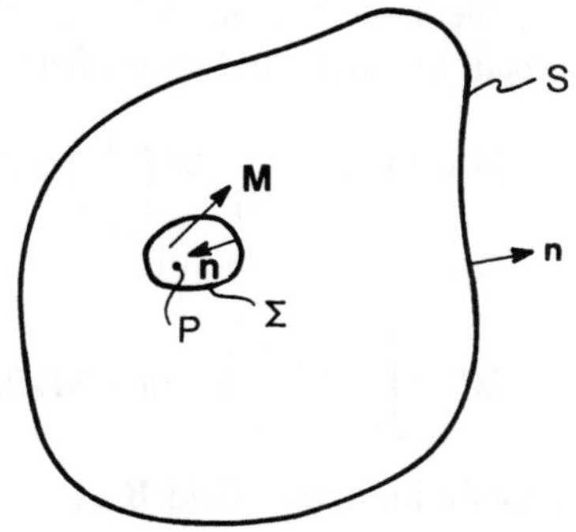

FIGURE 26. Entire magnetized region.

and we know that $\mathbf{A}' = \mathbf{A}$ in the limit $V'' \to 0$ ($V' \to V$). Application of the appropriate integral theorem to (7.2.3) yields[19]

$$\mathbf{A}' = -\int_S \frac{\mathbf{n} \times \mathbf{M}}{r_1} dS - \int_\Sigma \frac{\mathbf{n} \times \mathbf{M}}{r_1} dS + \int_{V'} \frac{\nabla^Q \times \mathbf{M}}{r_1} dV, \qquad (7.2.4)$$

which with (6.3.9) shows that the magnetic vector potential $\mathbf{A}$ due to the magnetization may be considered to be the same as that due to a volume distribution of current

$$\mathbf{J} = \nabla^Q \times \mathbf{M}, \qquad (7.2.5)$$

and a surface distribution of current

$$\bar{\mathbf{K}} = -\mathbf{n} \times \mathbf{M}, \qquad (7.2.6)$$

on the surface of the magnetized region S and the surface Σ excluding the domain V''. It is also clear from the expression for $\mathbf{A}$ in (7.2.4) that

$$\mathbf{A} = \lim_{V'' \to 0} \mathbf{A}', \qquad (7.2.7)$$

since it is clear that the integral over Σ vanishes if $|\mathbf{M}|$ is bounded and $r_1 \to 0$. However, as we know this will not be true for $\mathbf{B}'$.

The field $\mathbf{B}'$ is defined as

$$\mathbf{B}' = \nabla^P \times \mathbf{A}, \qquad (7.2.8)$$

hence from (7.2.4), we obtain

$$\mathbf{B}' = -\int_S \nabla^P \left(\frac{1}{r_1}\right) \times (\mathbf{n} \times \mathbf{M}) dS - \int_\Sigma \nabla^P \left(\frac{1}{r_1}\right) \times (\mathbf{n} \times \mathbf{M}) dS$$
$$+ \int_{V'} \nabla^P \left(\frac{1}{r_1}\right) \times (\nabla^Q \times \mathbf{M}) dV, \qquad (7.2.10)$$

from which with (6.1.10) it is clear that the integral over Σ gives a finite contribution[12] as $r_1 \to 0$, and the contribution depends on the shape of Σ.

In the same way and for the same reason as in the electric case in Sec. 3.2 we simply exclude this *local* magnetic induction field and define a $\mathbf{B}''$ as

$$\mathbf{B}'' = -\int_S \nabla^P\left(\frac{1}{r_1}\right) \times (\mathbf{n} \times \mathbf{M})\, dS + \int_{V'} \nabla^P\left(\frac{1}{r_1}\right) \times (\nabla^Q \times \mathbf{M})\, dV, \quad (7.2.11)$$

so that

$$\mathbf{B}' = \mathbf{B}'' - \int_\Sigma \nabla^P\left(\frac{1}{r_1}\right) \times (\mathbf{n} \times \mathbf{M}) dS, \quad (7.2.12)$$

and then define the magnetic induction field $\mathbf{B}$ as

$$\mathbf{B} = \lim_{V'' \to 0} \mathbf{B}'', \quad V' \to V, \quad (7.3.13)$$

from which it is clear that the magnetic induction field $\mathbf{B}$ does not determine the force on a current element $\mathbf{w}$ by the relation $\mathbf{w} \times \mathbf{B}$ in a magnetized region. In order to obtain the force on a current element, the local magnetic field $\mathbf{B}^L$ must be included. However, as we already know from the electric case, this fact affords no difficulty in a continuum theory because the mechanical force effect of this local field can readily be incorporated in the mechanical traction vector. In any even this is the field $\mathbf{B}$ that appears from this point on in magnetostatic (or electromagnetic) theory. Since it is clear that $\mathbf{A}'' = \mathbf{A}' = \mathbf{A}$ in the limit $V'' \to 0$, we still have

$$\mathbf{B} = \nabla^P \times \mathbf{A}, \quad (7.2.14)$$

within the magnetized region.

It should, of course, have been understood that in the foregoing treatment of magnetization the current density $\mathbf{J}$ has been assumed to vanish. Of course, if there is both current density $\mathbf{J}$ and magnetization $\mathbf{M}$, the resultant magnetic potential $\mathbf{A}$ and magnetic induction vector $\mathbf{B}$ may be obtained by superposition. Then, in the presence of both current density $\mathbf{J}$ and magnetization $\mathbf{M}$, from (6.3.10), (7.2.11) and (7.2.13) the expression for $\mathbf{B}$ *inside* or outside the magnetized region may be written in the form

$$\mathbf{B} = \int_V \nabla^P\left(\frac{1}{r_1}\right) \times \mathbf{J}\, dV + \int_V \nabla^P\left(\frac{1}{r_1}\right) \times (\nabla^Q \times \mathbf{M}) dV$$
$$- \int_S \nabla^P\left(\frac{1}{r_1}\right) \times (\mathbf{n} \times \mathbf{M}) dS. \quad (7.2.15)$$

Of course, as in (3.2.17) the dependence on the field point of each term is of the form in (6.1.10) which is particularly convenient for our purposes because it is perfectly clear from our previous work in Secs. 2.4, 3.2 and 6.3 that if $\mathbf{J}$ and $\nabla^P \times \mathbf{M}$ are bounded and integrable, $\mathbf{A}$ and $\mathbf{B}$ are continuous; and if $\mathbf{J}$ and $\nabla^Q \times \mathbf{M}$ and their first spatial derivatives are continuous, the first spatial derivatives of $\mathbf{B}$ and second spatial derivatives of $\mathbf{A}$ exist and are continuous at interior points.

At this point it should be noted that the expression for **A**, from which the expression for **B** in (7.2.15) can be obtained, may be written

$$\mathbf{A} = \int_V \frac{1}{r_1}(\mathbf{J} + \nabla^Q \times \mathbf{M})dV - \int_S \frac{\mathbf{n} \times \mathbf{M}}{r_1} dS. \tag{7.2.16}$$

In this expression the term

$$-\int_\Sigma \frac{\mathbf{n} \times \mathbf{M}}{r_1} dS, \tag{7.2.17}$$

which vanishes as $\Sigma \to 0$ has been omitted since it yields the local portion of the magnetic induction vector $\mathbf{B}^L$, which does not vanish as $\Sigma \to 0$ and has been omitted from the definition of **B**. We now wish to determine the expression for $\nabla^P \cdot \mathbf{A}$ at P for use in the next section. To this end we use the cavity definition of **A** in (7.2.7), which is equivalent to the use of (7.2.16), because its use is necessary when terms contained in $\nabla^P \mathbf{A}$ are required. Now, from (7.2.16) and the appropriate integral theorem[19] along with (7.2.7), we may write

$$\mathbf{A} = \lim_{V'' \to 0} \mathbf{A}' = \lim_{V'' \to 0} \int_{V'} \left[\frac{\mathbf{J}}{r_1} + \nabla^P\left(\frac{1}{r_1}\right) \times \mathbf{M}\right] dV. \tag{7.2.18}$$

From (7.2.18) we obtain

$$\nabla^P \cdot \mathbf{A} = \lim_{V'' \to 0} \int_{V'} \left[\nabla^P\left(\frac{1}{r_1}\right) \cdot \mathbf{J} + \nabla^P \cdot \nabla^P\left(\frac{1}{r_1}\right) \times \mathbf{M}\right] dV, \tag{7.2.19}$$

which may be written

$$\nabla^P \cdot \mathbf{A} = -\lim_{V'' \to 0} \int_{V'} \nabla^Q\left(\frac{1}{r_1}\right) \cdot \mathbf{J}\, dV + \lim_{V'' \to 0} \int_{V'} \nabla^P \cdot \nabla^P \times \left(\frac{\mathbf{M}}{r_1}\right) dV, \tag{7.2.20}$$

since $\mathbf{M} = \mathbf{M}(Q)$. Thus, the last term vanishes and we have

$$\nabla^P \cdot \mathbf{A} = -\lim_{V'' \to 0} \int_{V'} \left[\nabla^Q \cdot \left(\frac{\mathbf{J}}{r_1}\right) - \frac{\nabla^Q \cdot \mathbf{J}}{r_1}\right] dV, \tag{7.2.21}$$

which, with (6.3.23) for *steady* currents and the divergence theorem, enables us to write

$$\nabla^P \cdot \mathbf{A} = -\int_S \frac{\mathbf{n} \cdot \mathbf{J}}{r_1} dS - \lim_{\Sigma \to 0} \int_\Sigma \frac{\mathbf{n} \cdot \mathbf{J}}{r_1} dS, \tag{7.2.22}$$

where S is the surface surrounding the entire magnetized region carrying current and Σ is the closed surface surrounding V''. Now, since

$$\lim_{\Sigma \to 0} \int_\Sigma \frac{\mathbf{n} \cdot \mathbf{J}}{r_1} dS = 0, \tag{7.2.23}$$

we have

$$\nabla^P \cdot \mathbf{A} = -\int_S \frac{\mathbf{n} \cdot \mathbf{J}}{r_1} dS. \tag{7.2.24}$$

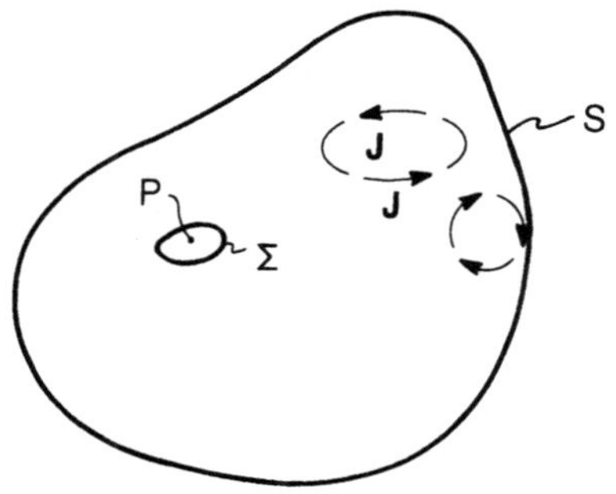

FIGURE 27. Entire magnetized region carrying current.

Since S encloses the *entire* region carrying current as shown in Figure 27, no current $\mathbf{J}$ can cross S at *any* point, and on S we have

$$\mathbf{n}\cdot\mathbf{J} = 0, \qquad (7.2.25)$$

which with (7.2.24) yields

$$\nabla^P\cdot\mathbf{A} = 0. \qquad (7.2.26)$$

7.3 Field Equation of Magnetostatics

The expression for $\nabla^P\mathbf{A}$, in which the local field terms $(\nabla^P\mathbf{A})^L$ have been excluded, may be obtained from (7.2.16) in the form

$$\nabla^P\mathbf{A} = \int_V \nabla^P\left(\frac{1}{r_1}\right)(\mathbf{J}\times\nabla^Q\times\mathbf{M})dV - \int_S \nabla^P\left(\frac{1}{r_1}\right)\mathbf{n}\times\mathbf{M}\,dS. \qquad (7.3.1)$$

Now, consider the entire current carrying and magnetized region, which is enclosed within S_0 shown in Figure 28, and consider any surface $\hat{S}$ within which $\mathbf{J}, \nabla^Q\times\mathbf{M}$ and their gradients are continuous. Integrate the normal components of $\nabla^P\mathbf{A}$ over $\hat{S}$ to form the vector

$$\int_{\hat{S}} \mathbf{n}\cdot\nabla^P\mathbf{A}\,dS, \qquad (7.3.2)$$

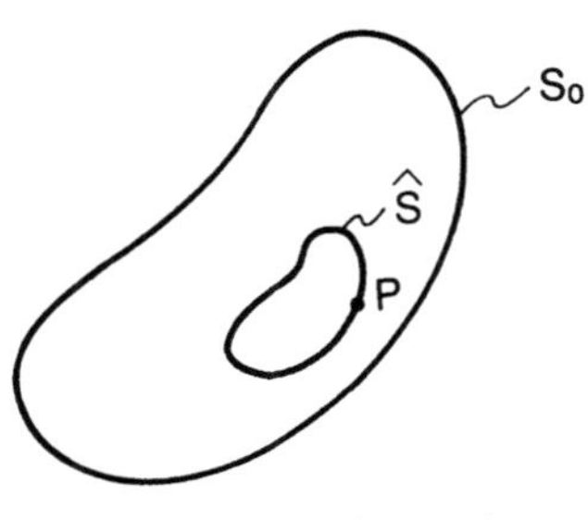

FIGURE 28. Entire current carrying and magnetized region.

which, since **A** is the vector potential due to all current and magnetization contained within S_0, with (7.3.1) enables us to write

$$\int_{\hat{S}} \mathbf{n}\cdot\nabla^P \mathbf{A}\, dS = \int_{\hat{S}} \mathbf{n}\cdot\left[\int_{V_0} \nabla^P\left(\frac{1}{r_1}\right)(\mathbf{J} + \nabla^Q \times \mathbf{M})dV - \int_{S_0} \nabla^P\left(\frac{1}{r_1}\right)\mathbf{n}\times\mathbf{M}\, dS_0\right]dS. \tag{7.3.3}$$

Interchanging the order of integration and substituting from (6.1.10), we obtain

$$\int_{\hat{S}} \mathbf{n}\cdot\nabla^P \mathbf{A}\, dS = -\int_{V_0} (\mathbf{J} + \nabla^Q \times \mathbf{M}) \int_{\hat{S}} \frac{\mathbf{n}\cdot r_1}{r_1^3}\, dS\, dV + \int_{S_0} \mathbf{n}\times\mathbf{M} \int_{\hat{S}} \frac{\mathbf{n}\cdot\mathbf{r}_1}{r_1^3}\, dS\, dS_0. \tag{7.3.4}$$

Integrating the right-hand side of (7.3.4) over the *closed* surface $\hat{S}$ and employing (2.3.7)–(2.3.9) we obtain

$$\int_{\hat{S}} \mathbf{n}\cdot\nabla^P \mathbf{A}\, dS = -4\pi \int_{\hat{V}} (\mathbf{J} + \nabla^Q \times \mathbf{M})\, dV, \tag{7.3.5}$$

since, as shown in Sec. 2.3, for a closed surface $\hat{S}$, the solid angle for any point inside is 4π, and all of S_0 and $V_0 - \hat{V}$ are outside $\hat{S}$. Since the integration variables are dummy variables and we restrict **J** and $\nabla^Q \times \mathbf{M}$ to satisfy the conditions of the divergence theorem,[20] we obtain

$$\int_{\hat{V}} (\nabla^2 \mathbf{A} + 4\pi\nabla\times\mathbf{M} + 4\pi\mathbf{J})dV = 0, \tag{7.3.6}$$

for *any* volume $\hat{V}$. Hence, we have

$$\nabla^2 \mathbf{A} = -4\pi\nabla\times\mathbf{M} - 4\pi\mathbf{J}, \tag{7.3.7}$$

at all points of the magnetized region carrying current. Equation (7.3.7) is one form of the differential equation of magnetostatics.

By virtue of the well-known identity

$$\nabla\times\nabla\times\mathbf{A} = \nabla\nabla\cdot\mathbf{A} - \nabla^2\mathbf{A}, \tag{7.3.8}$$

and (7.2.26), Eq. (7.3.7) may be written in the form

$$\nabla\times\nabla\times\mathbf{A} = 4\pi\mathbf{J} + 4\pi\nabla\times\mathbf{M}, \tag{7.3.9}$$

which with (7.2.14) may be written

$$\nabla\times(\mathbf{B} - 4\pi\mathbf{M}) = 4\pi\mathbf{J}. \tag{7.3.10}$$

The vector quantity $\mathbf{B} - 4\pi\mathbf{M}$ is a very important magnetic quantity called the magnetic field and is given the symbol **H**. Thus, we define

$$\mathbf{H} = \mathbf{B} - 4\pi\mathbf{M}, \tag{7.3.11}$$

in Gaussian units, which enables us to write the differential equations of magnetostatics in the form

$$\nabla \times \mathbf{H} = 4\pi \mathbf{J}, \tag{7.3.12}$$

$$\mathbf{H} = \mathbf{B} - 4\pi \mathbf{M}, \tag{7.3.13}$$

$$\nabla \cdot \mathbf{B} = 0. \tag{7.3.14}$$

These equations in this form are quite general but cannot be solved unless there is either a relation between **M** and **B**, or **M** is given.

7.4 Linearized Magnetostatics

In ordinary elementary magnetostatics a linear relation between **M** and **H** is assumed, i.e.,

$$\mathbf{M} = \psi \mathbf{H}, \tag{7.4.1}$$

for an isotropic material and

$$M_i = \psi_{ij} H_j, \tag{7.4.2}$$

for an anisotropic material, where ψ_{ij} is not yet symmetric. The symbol ψ_{ij} is called the magnetic susceptibility tensor. When these linear relationships exist, we have

$$B_i = H_i + 4\pi M_i = (\delta_{ij} + 4\pi\psi_{ij})H_j = \mu_{ij} H_j, \tag{7.4.3}$$

where

$$\mu_{ij} = \delta_{ij} + 4\pi\psi_{ij}, \tag{7.4.4}$$

is the magnetic permeability tensor. For an isotropic material we have

$$\mu_{ij} = \delta_{ij}(1 + 4\pi\psi) = \mu\delta_{ij}, \tag{7.4.5}$$

so that we may write

$$\mathbf{B} = \mu \mathbf{H}, \tag{7.4.6}$$

where μ is called the permeability constant. All of these linear relationships are for paramagnetic and diamagnetic media, but not for ferromagnetic media. We will not say much about ferromagnetic media here. The difference between paramagnetic and diamagnetic materials is that in diamagnetic materials the susceptibility ψ is negative and in paramagnetic materials positive.

Now, in an isotropic linear magnetostatic problem, from (7.4.6), (7.3.12), (7.2.14) and (7.3.8) we have

$$\nabla^2 \mathbf{A} = -4\pi\mu \mathbf{J}, \tag{7.4.7}$$

as the vector differential equation valid in each region with different μ and **J**. Thus it is clear that in magnetostatics we have a *vector* field **A**, which satisfies a *vector* Poisson equation, in contrast with electrostatics, in which we have a *scalar* field φ satisfying a scalar Poisson equation. Now, in order to solve a boundary value problem in magnetostatics, we must determine the magnetic boundary conditions. However, just as in the electrostatic case, before we obtain the boundary conditions, we observe that in actuality we have no particular interest in the linear constitutive equations we have written in this section. Hence, from our point of view we need the boundary conditions that go along with (7.3.12)–(7.3.14).

7.5 Boundary Conditions of Magnetostatics

The differential equations of magnetostatics are given by (7.3.12)–(7.3.14), and we note that (7.3.14) can be written in the equivalent form

$$\mathbf{B} = \nabla \times \mathbf{A}. \tag{7.5.1}$$

As was stated in the electrostatic case, the truly fundamental equations must always be integral statements from which the governing differential equations may be derived when suitable differentiability and continuity conditions are satisfied. Consequently, as in Sec. 3.5, we obtain the integral statements corresponding to the pertinent differential statements, and then *take* the integral statement to hold even when the differential statements do not, as across surfaces of discontinuity. Clearly, the integral forms corresponding to (7.3.12) and (7.3.14), respectively, are

$$\oint_C \mathbf{H} \cdot d\mathbf{r} = 4\pi \int_s \mathbf{n} \cdot \mathbf{J}\, ds, \tag{7.5.2}$$

$$\int_S \mathbf{n} \cdot \mathbf{B}\, dS = 0, \tag{7.5.3}$$

where C is *any* closed curve enclosing an open area s and S is *any* closed surface. These are the appropriate integral forms because, if $\mathbf{H}$ and $\mathbf{B}$ are sufficiently differentiable for the divergence theorem and Stokes theorem to be valid, they yield (7.3.12) and (7.3.14), respectively.

Since the integral forms are *taken* to be valid across surfaces of discontinuity, we may use them to obtain the boundary conditions in the same way as in electrostatics. However, whereas the electric charge density ρ was allowed to be singular at a surface and yield a surface charge density σ, the electric current density $\mathbf{J}$ is *not* allowed to be singular at a surface, and, consequently, there can be no surface current density $\mathbf{K}$. Then it is perfectly clear from our work with the electrostatic boundary conditions in Sec. 3.5 that the magnetostatic boundary (jump) conditions obtained from (7.5.2) and (7.5.3) take the respective forms

$$\mathbf{n} \times [\mathbf{H}] = 0, \tag{7.5.4}$$

$$\mathbf{n} \cdot [\mathbf{B}] = 0. \tag{7.5.5}$$

Equations (7.5.4) and (7.5.5), respectively, say that the tangential components of **H** and normal components of **B** are continuous across any surface of discontinuity. If a surface current density **K** had been allowed to exist it would have appeared on the right-hand side of (7.5.4). However, we exclude this possibility as unphysical inasmuch as it will not occur in any situation encountered in this treatment.

Since it was shown in Sec. 3.5 that the electric scalar potential φ was continuous across surfaces of discontinuity, we now show that the magnetic vector potential **A** is continuous across surfaces of discontinuity. To this end we substitute from (7.2.14) into (7.5.3) to obtain

$$\int_S \mathbf{n}\cdot\nabla\times\mathbf{A}\,dS = 0, \tag{7.5.6}$$

which with Stokes theorem yields

$$\oint_C \mathbf{A}\cdot d\mathbf{r} = 0, \tag{7.5.7}$$

from which we readily obtain the jump conditions

$$\mathbf{n}\times[\mathbf{A}] = 0. \tag{7.5.8}$$

From (7.2.26) with the aid of the divergence theorem we may write

$$\int_S \mathbf{n}\cdot\mathbf{A}\,dS = 0, \tag{7.5.9}$$

for an arbitrary closed surface. Then from (7.5.9) we readily obtain the jump condition

$$\mathbf{n}\cdot[\mathbf{A}] = 0. \tag{7.5.10}$$

Equations (7.5.8) and (7.5.10) permit us to write

$$[\mathbf{A}] = 0, \tag{7.5.11}$$

which shows that **A** is continuous across surfaces of discontinuity.

CHAPTER 8

Forces and Torques Exerted by the Magnetic Induction Field on Magnetized Matter Carrying Current

8.1 Magnetic Body Forces

It is quite clear from the material in Chapter 6 that by definition the force exerted by the magnetic induction field **B** on a differential element of stationary volume containing a current density **J** is given by

$$d\mathbf{F} = \mathbf{J} \times \mathbf{B}\, dV. \tag{8.1.1}$$

If no magnetization is present at the point in question this body force represents the total magnetic force exerted on $\mathbf{J}\,dV$. However, if magnetization is present, as we already know, a portion of the magnetic force on **J** at Q is due to a local surface effect, which depends on the shape of the volume element, and in a continuum theory can readily be included as a portion of the mechanical traction vector. Nevertheless, $\mathbf{J} \times \mathbf{B}\, dV$ is still the body force exerted by the magnetic induction vector **B** on $\mathbf{J}\,dV$, since the local magnetic force, which depends on the shape of the surface surrounding the volume element, will be included with the mechanical traction.

Now, we wish to know the expression for the body force exerted by the magnetic induction field **B** on magnetization **M** at a region dV. To this end we consider the force exerted by the magnetic induction field **B** on a planar current loop of area S_0, steady current I and normal to plane surface **n**, and take the limit of the expression as $S_0 \to 0, I \to \infty$ with **n** fixed while the product $S_0 I \to m$. From Figure 29 and the material in Chapter 6 it is clear that the force is given by

$$\mathbf{f} = I \oint_C d\mathbf{s} \times \mathbf{B}, \tag{8.1.2}$$

which, with the aid of the integral theorem,[21]

$$\oint_C d\mathbf{r} \times \mathbf{G} = \int_S (\mathbf{n} \times \nabla) \times \mathbf{G}\, dS, \tag{8.1.3}$$

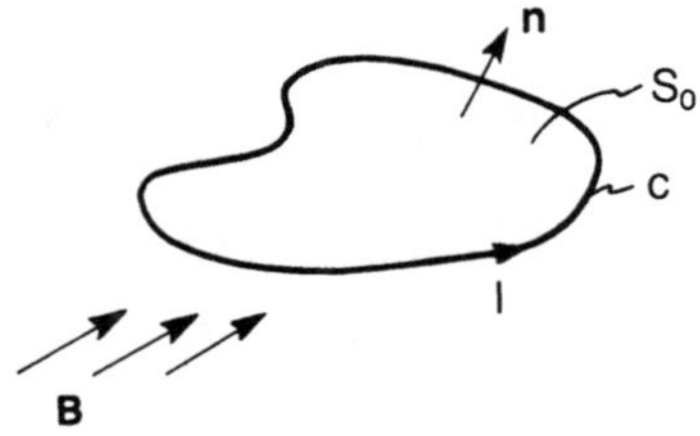

FIGURE 29. Force exerted by magnetic induction field on a point magnetic dipole moment.

may be written in the form

$$\mathbf{f} = I \int_{S_0} (\mathbf{n} \times \nabla) \times \mathbf{B}\, dS. \tag{8.1.4}$$

In indicial notation the integral in (8.1.4) takes the form

$$\begin{aligned}[(\mathbf{n} \times \nabla) \times \mathbf{B}]_l &= e_{ijk} n_j \frac{\partial}{\partial x_k} e_{lim} B_m \\ &= e_{ijk} e_{iml} n_j B_{m,k}, \end{aligned} \tag{8.1.5}$$

where e_{ijk} denotes the skew symmetric tensor or Levi-Civita symbol.[22] Using a well-known tensor identity,[22] we have

$$\begin{aligned}[(\mathbf{n} \times \nabla) \times \mathbf{B}]_l &= (\delta_{jm}\delta_{kl} - \delta_{jl}\delta_{km}) n_j B_{m,k} \\ &= n_m B_{m,l} - n_l B_{k,k}, \end{aligned} \tag{8.1.6}$$

which, with (7.3.14), yields

$$[(\mathbf{n} \times \nabla) \times \mathbf{B}]_l = n_m B_{m,l}. \tag{8.1.7a}$$

The vector form corresponding to (8.1.7a) may readily be constructed from it thus

$$(\mathbf{n} \times \nabla) \times \mathbf{B} = \mathbf{n}_m e_l \frac{\partial}{\partial x_l} B_m = n_k \mathbf{e}_k \cdot \mathbf{e}_m \frac{\partial}{\partial x_l} B_m \mathbf{e}_l = \mathbf{n} \cdot (\nabla \mathbf{B})^T = \mathbf{n} \cdot (\mathbf{B}\nabla). \tag{8.1.7b}$$

Substituting from (8.1.7b) into (8.1.4), we obtain

$$\mathbf{f} = \mathbf{I} \int_{S_0} \mathbf{n} \cdot (\mathbf{B}\nabla) dS, \tag{8.1.8}$$

and now we take the limit as $S_0 \to 0$ and $I \to \infty$ with $\mathbf{n}$ fixed and $IS_0 \to m$. Under these circumstances, by the mean value theorem or Taylor's theorem we can take $\mathbf{B}\nabla$ out of the integral and write

$$\mathbf{f} = \lim_{\substack{S_0 \to 0 \\ I \to \infty \\ \mathbf{n}\ \text{fixed}}} I\mathbf{n}S_0 \cdot (\mathbf{B}\nabla) = \mathbf{m} \cdot (\mathbf{B}\nabla). \tag{8.1.9}$$

Equation (8.1.9) gives the expression for the force exerted by an external magnetic induction field **B** on a single magnetic dipole moment **m**. Clearly, we may repeat this for a number of discrete magnetic dipole moments at distinct points (n) just as we did in Sec. 3.1 to obtain

$$\mathbf{f} = \sum_{n=1}^{N} \mathbf{m}^{(n)} \cdot (\mathbf{B}^{(n)} \nabla^{(n)}). \tag{8.1.10}$$

Now we note that for the microscopic region of interest, which enables us to define the macroscopic magnetization density **M**, the gradient of the macroscopic magnetic induction field has one value. Consequently, for a region of that size Eq. (8.1.10) may be written in the form

$$\mathbf{f} = \sum_{n=1}^{K} \mathbf{m}^{(n)} \cdot (\mathbf{B} \nabla). \tag{8.1.11}$$

Then by means of the usual reasoning to go from the discrete to the continuous, we may write

$$d\mathbf{F} = \mathbf{M} \cdot \mathbf{B} \nabla. \tag{8.1.12}$$

Clearly then, from (8.1.1) and (8.1.12) the expression for the force exerted by the magnetic induction on continuous distributions of current density **J** and magnetization **M** in a differential element of volume dV may be written in the form

$$d\mathbf{F} = [\mathbf{J} \times \mathbf{B} + \mathbf{M} \cdot (\mathbf{B} \nabla)] dV. \tag{8.1.13}$$

If we integrate (8.1.13) over some portion of or an entire body, we obtain

$$\mathbf{F} = \int_V [\mathbf{J} \times \mathbf{B} + \mathbf{M} \cdot (\mathbf{B} \nabla)] dV = \int_V \mathbf{f}\, dV, \tag{8.1.14}$$

where **F** is the force exerted by **B** on either some portion of or an entire body and we note that **B** in (8.1.14) arises from sources both inside and outside the region denoted V. Thus we have shown that the expression for the body force density **f** exerted by **B** on magnetized matter carrying current may be written in the form

$$\mathbf{f} = \mathbf{J} \times \mathbf{B} + \mathbf{M} \cdot (\mathbf{B} \nabla). \tag{8.1.15}$$

8.2 Maxwell Magnetostatic Stress Tensor

Now that we have the expression (8.1.15) for the body force exerted by the magnetic induction field on magnetized continua carrying current, we substitute from (7.3.12) into (8.1.15) to obtain

$$\mathbf{f} = \frac{1}{4\pi} (\nabla \times \mathbf{H}) \times \mathbf{B} + \mathbf{M} \cdot (\mathbf{B} \nabla), \tag{8.2.1a}$$

or in Cartesian component form

$$f_l = \frac{1}{4\pi} e_{ijk} H_{k,j} e_{\text{lim}} B_m + M_k B_{k,l}, \tag{8.2.1b}$$

which shows, as we expected, that the magnetic body force term is automatically nonlinear in the magnetic field variables, and, hence, of course, any coupled magnetomechanical system is inherently nonlinear. Employing the well-known identity we used in going from (8.1.5) to (8.1.6) in (8.2.1b), we obtain

$$f_l = \frac{1}{4\pi}(H_{l,m} B_m - H_{k,l} B_k) + M_k B_{k,l}, \tag{8.2.2}$$

which with (7.3.14) and (7.3.13) enables us to write

$$f_l = \frac{1}{4\pi}(H_l B_m)_{,m} - \frac{1}{4\pi} H_k H_{k,l} + 4\pi M_k M_{k,l}, \tag{8.2.3}$$

which clearly may be written in the form

$$f_l = \frac{1}{4\pi}[H_l B_m + \tfrac{1}{2}(16\pi^2 M_k M_k - H_k H_k)\delta_{lm}]_{,m}, \tag{8.2.4}$$

and this last expression within the brackets is clearly a second rank tensor and is called the Maxwell magnetostatic stress tensor. Hence, we may write

$$T^M_{ml} = \frac{1}{4\pi}[B_m H_l + \tfrac{1}{2}(16\pi^2 M_k M_k - H_k H_k)\delta_{ml}], \tag{8.2.5}$$

for the Maxwell magnetostatic stress tensor, which enables (8.2.4) to be written in the form

$$f_l = T^M_{ml,m}, \tag{8.2.6}$$

or in invariant dyadic form (8.2.5) may be written

$$\mathbf{T}^M = \frac{1}{4\pi}[\mathbf{BH} + \tfrac{1}{2}(16\pi^2 \mathbf{M}\cdot\mathbf{M} - \mathbf{H}\cdot\mathbf{H})\mathbf{I}], \tag{8.2.7}$$

and (8.2.6) may be written

$$\mathbf{f} = \nabla\cdot\mathbf{T}^M. \tag{8.2.8}$$

Consequently, the force **F** exerted by **B** on a volume of magnetized matter carrying current may be written

$$\mathbf{F} = \int_V \mathbf{f}\, dV = \int_V \nabla\cdot\mathbf{T}^M\, dV, \tag{8.2.9}$$

and with the aid of the divergence theorem, we obtain

$$\mathbf{F} = \int_S \mathbf{n}\cdot\mathbf{T}^M\, dS. \tag{8.2.10}$$

Thus we have shown that the force exerted by the magnetic induction vector $\mathbf{B}$ on an element of matter may be represented as an integral of the Maxwell magnetostatic traction force over the surface S surrounding V. As in the electric case, this is very important because, as usual, integral statements are more fundamental than differential statements because they can be *taken* to exist even when the differential statements do not. Naturally, we take the Maxwell magnetostatic stress tensor $\mathbf{T}^M$ to represent the mechanical force effect of $\mathbf{B}$ as a force acting over the surface of an arbitrary region, in general, i.e., even when the field vectors are discontinuous and $\nabla \cdot \mathbf{T}^M$ does not exist. Of course, when the field vectors are suitably differentiable, $\nabla \cdot \mathbf{T}^M$ exists and we obtain (8.2.8) from (8.2.10). However, as we already know the field vector may not be continuous at material surfaces of discontinuity, and at such places, $\mathbf{T}^M$ must be used. Thus, it is quite clear that the expression for $\mathbf{T}^M$ will be very important in its effect on the mechanical boundary conditions at surfaces of discontinuity.

8.3 Magnetic Torques

At this point it should be noted that in general $\mathbf{T}^M$ is not symmetric. In fact, from (8.2.7) and (7.3.13) its antisymmetric part is given by

$$(\mathbf{T}^M)^A = \tfrac{1}{2}(\mathbf{MB} - \mathbf{BM}), \tag{8.3.1}$$

and $(\mathbf{T}^M)^A$ will not be zero if $\mathbf{M}$ and $\mathbf{B}$ are not colinear. This means in essence that the magnetic induction vector $\mathbf{B}$ can exert a couple on the magnetization $\mathbf{M}$, and that, consequently, the Cauchy mechanical stress tensor cannot be expected to be symmetric in general when coupling with the magnetic field exists.

Let us now obtain the expression for the couple exerted by $\mathbf{B}$ on $\mathbf{M}$ directly from current loop considerations. To this end we again consider the diagram shown in Figure 29 and evaluate the couple exerted by the magnetic induction field $\mathbf{B}$ on a planar current loop enclosing an area S_0, by taking moments about a point 0 within S_0, and taking the usual limit. Clearly, the expression for the couple may be written

$$\mathbf{L} = I \oint_C \mathbf{r} \times (d\mathbf{s} \times \mathbf{B}), \tag{8.3.2}$$

which in indicial notation takes the form

$$L_j = I \oint_C e_{jkl} e_{lmn} x_k B_n ds_m, \tag{8.3.3}$$

which may be written

$$L_j = I \oint_C g_{mj} ds_m, \tag{8.3.4}$$

where

$$g_{mj} = e_{jkl}e_{lmn}x_k B_n. \tag{8.3.5}$$

We now employ Stokes' theorem for tensor point functions, which may be written[23]

$$\int_S \mathbf{n}\cdot\nabla\times\mathbf{f}\,dS = \oint_C d\mathbf{r}\cdot\mathbf{f}, \tag{8.3.6a}$$

and in indicial notation takes the form

$$\int_S n_i e_{irs} f_{st,r}\,dS = \oint_C dx_n f_{nt}. \tag{8.3.6b}$$

Application of (8.3.6b) to (8.3.4) enables us to write

$$L_j = I\oint_C g_{mj}\,ds_m = I\int_S n_i e_{irs} g_{sj,r}\,dS, \tag{8.3.7}$$

which with (8.3.5) yields

$$L_j = I\int_S n_i e_{iks} e_{jkl} e_{lsn} B_n\,dS + I\int_S n_i e_{irs} e_{jkl} e_{lsn} x_k B_{n,r}\,dS. \tag{8.3.8}$$

Now, with the identity used in (8.1.6) the first integral may be written

$$I\int_S n_i(\delta_{sl}\delta_{ij} - \delta_{sj}\delta_{il})e_{lsn} B_n\,dS = I\int_S n_l e_{jln} B_n\,dS. \tag{8.3.9}$$

Hence, Eq. (8.3.8) takes the form

$$L_j = I\int_S n_l e_{jln} B_n\,dS + I\int_S n_i e_{irs} e_{jkl} e_{lsn} x_k B_{n,r}\,dS, \tag{8.3.10}$$

and, now, if we take the limit as $S_0 \to 0$ and $I \to \infty$ with $\mathbf{n}$ fixed and $IS_0 \to m$, we obtain

$$L_j = \lim_{\substack{S_0\to 0\\ I\to\infty}} n_l I S_0 e_{jln} B_n + \lim_{\substack{S_0\to 0\\ I\to\infty}} I n_i e_{irs} e_{jkl} e_{lsn} B_{n,r}\int_{S_0} x_k\,dS, \tag{8.3.11}$$

but

$$\lim_{\substack{S_0\to 0\\ I\to\infty}} I\int_{S_0} x_k\,dS < \lim_{\substack{S_0\to 0\\ I\to\infty}} IS_0(x_k)^{\max} = m\lim_{S_0\to 0}(x_k)^{\max} \to 0. \tag{8.3.12}$$

Thus, in the limit, we have

$$L_j = n_l m e_{jln} B_n, \tag{8.3.13}$$

or in vector notation

$$\mathbf{L} = \mathbf{m}\times\mathbf{B}, \tag{8.3.14}$$

for a point magnetic dipole. Then by means of an argument similar to the one employed in going from (8.1.9) to (8.1.14), we obtain

$$\mathbf{L} = \int_V \mathbf{M} \times \mathbf{B}\, dV = \int_V \mathbf{l}\, dV. \tag{8.3.15}$$

Clearly, in component form we have

$$l_i = e_{ijk} M_j B_k, \tag{8.3.16}$$

where l_i is an axial vector representing the torque density. As in the electric case it can be shown in the usual way that this couple is equivalent to the antisymmetric portion of the Maxwell magnetostatic stress tensor, i.e., by writing the antisymmetric stress tensor representation of the axial vector thus

$$(T^M_{lm})^A = \tfrac{1}{2} e_{ilm} l_i = \tfrac{1}{2} e_{ilm} e_{ijk} M_j B_k = \tfrac{1}{2}(M_l B_m - B_l M_m), \tag{8.3.17}$$

which is the component form of (8.3.1).

CHAPTER 9

Magnetostatic Energy

9.1 Energy Resulting from Distributions of Current

This brings us to a consideration of magnetostatic energy. We begin by considering a collection of elementary *closed* current loops $C^{(n)}$, with each loop having a *steady* current $I^{(n)}$. We first bring in the first current loop $C^{(1)}$ with steady current $I^{(1)}$ to its location and at this point no magnetic field energy exists because we exclude the self energy of the loop, which is singular. Then we bring in a second loop $C^{(2)}$ with steady current $I^{(2)}$ to its position and determine the interaction energy of the two loops, excluding the singular self energy of both. We have already calculated this energy in Chapter 6 and from Eq. (6.2.12) we have

$$u_2 = I^{(1)}I^{(2)}\oint_{C^{(1)}}\oint_{C^{(2)}}\frac{d\mathbf{s}^{(1)}\cdot d\mathbf{s}^{(2)}}{r^{(12)}}. \tag{9.1.1}$$

Now, we bring a third loop $C^{(3)}$ with steady current $I^{(3)}$ to its position and determine the interaction energy of this loop with the other two, excluding the singular self-energy of all three. Then we bring a fourth loop to its location and determine the interaction energy with the other three, excluding all singular self-energies and so on. Clearly then, we can write for the interaction energy of each loop with all the other loops the relations

$$\begin{aligned}
u_1 &= 0,\\
u_2 &= I^{(1)}I^{(2)}\oint_{C^{(1)}}\oint_{C^{(2)}}\frac{d\mathbf{s}^{(1)}\cdot d\mathbf{s}^{(2)}}{r^{(12)}},\\
u_3 &= I^{(1)}I^{(3)}\oint_{C^{(1)}}\oint_{C^{(3)}}\frac{d\mathbf{s}^{(1)}\cdot d\mathbf{s}^{(3)}}{r^{13}} + I^{(2)}I^{(3)}\oint_{C^{(2)}}\oint_{C^{(3)}}\frac{d\mathbf{s}^{(2)}\cdot d\mathbf{s}^{(3)}}{r^{23}},\\
u_4 &= \cdots,\\
u_5 &= \cdots \text{and so on to } u_N,
\end{aligned} \tag{9.1.2}$$

where u_n represents the energy of interaction of each loop with all the other loops. Then the total magnetic field energy $\mathscr{U}$ may be written

$$\mathscr{U} = u_1 + u_2 + u_3 + \cdots + u_N, \tag{9.1.3}$$

which clearly may be written in the entirely equivalent form

$$\begin{aligned}\mathscr{U} = {} & \tfrac{1}{2}I^{(1)}\left[I^{(2)}\oint_{C(1)}\oint_{C(2)}\frac{d\mathbf{s}^{(1)}\cdot d\mathbf{s}^{(2)}}{r^{12}} + I^{(3)}\oint_{C(1)}\oint_{C(3)}\frac{d\mathbf{s}^{(1)}\cdot d\mathbf{s}^{(3)}}{r^{12}}\right.\\ & \left. + \cdots + I^{(N)}\oint_{C(1)}\oint_{C(N)}\frac{d\mathbf{s}^{(1)}\cdot d\mathbf{s}^{(N)}}{r^{1N}}\right]\\ & + \tfrac{1}{2}I^{(2)}\oint_{C(2)}\left[I^{(1)}\oint_{C(1)}\frac{d\mathbf{s}^{(2)}\cdot d\mathbf{s}^{(2)}}{r^{21}} + I^{(3)}\oint_{C(3)}\frac{d\mathbf{s}^{(2)}\cdot d\mathbf{s}^{(3)}}{r^{23}}\right.\\ & \left. + \cdots + I^{(N)}\int_{C(N)}\frac{d\mathbf{s}^{(2)}\cdot d\mathbf{s}^{(N)}}{r^{2N}}\right] + \cdots\\ & + \tfrac{1}{2}I^{(N)}\oint_{C(N)}\left[I^{(1)}\oint_{C(1)}\frac{d\mathbf{s}^{(N)}\cdot d\mathbf{s}^{(1)}}{r^{N1}} + I^{(2)}\oint_{C(2)}\frac{d\mathbf{s}^{(N)}\cdot d\mathbf{s}^{(2)}}{r^{N2}}\right.\\ & \left. + \cdots + I^{(N-1)}\oint_{C(N-1)}\frac{d\mathbf{s}^{(N)}\cdot d\mathbf{s}^{(N-1)}}{r^{N(N-1)}}\right],\end{aligned} \tag{9.1.4}$$

and using the definitions in Eq. (6.2.5) this may be written in the more convenient form

$$\begin{aligned}\mathscr{U} = {} & \frac{1}{2}\oint_{C(1)}\mathbf{w}^{(1)}\cdot\left[\oint_{C(2)}\frac{\mathbf{w}^{(2)}}{r^{12}} + \oint_{C(3)}\frac{\mathbf{w}^{(3)}}{r^{13}} + \cdots + \oint_{C(N)}\frac{\mathbf{w}^{(N)}}{r^{1N}}\right]\\ & + \tfrac{1}{2}\mathbf{w}^{(2)}\cdot\left[\oint_{C(1)}\frac{\mathbf{w}^{(1)}}{r^{21}} + \oint_{C(3)}\frac{\mathbf{w}^{(3)}}{r^{23}} + \cdots + \oint_{C(N)}\frac{\mathbf{w}^{(N)}}{r^{2N}}\right] + \cdots\\ & + \frac{1}{2}\oint_{C(N)}\mathbf{w}^{(N)}\cdot\left[\oint_{C(1)}\frac{\mathbf{w}^{(1)}}{r^{N1}} + \oint_{C(2)}\frac{\mathbf{w}^{(2)}}{r^{N2}} + \cdots + \oint_{C(N-1)}\frac{\mathbf{w}^{(N-1)}}{r^{N(N-1)}}\right].\end{aligned} \tag{9.1.5}$$

It is clear from (6.2.19) that this expression may be written in the form

$$\mathscr{U} = \frac{1}{2}\sum_{n=1}^{N}\oint_{C(n)}\mathbf{w}^{(n)}\cdot\mathbf{A}^{(n)}, \tag{9.1.6}$$

where $\mathbf{A}^{(n)}$ is the magnetic vector potential at *each point* of the nth current loop due to all the *other* current loops and excluding any interaction of current elements of the nth loop with *different* current elements of the *same* nth loop. This excluded energy due to the interaction of *different* current elements of the *same* current loop is the portion of the self energy of that current loop which is not singular.

We would like to, and indeed should, include this nonsingular contribution of the self energy of each current loop to the magnetostatic field energy.

However, we have defined the energy of a magnetostatic system consisting of current loops only and as being the interaction energy *between entire* closed current loops because the spatial gradient of this defined energy is equal to the *external* force exerted by the *other* closed current loops on the considered *entire* closed current loop. We now *generalize* our theory so as to include the *finite* portion of the self energy of *each* current loop in the magnetic field energy. We do this by assuming that the expression we have obtained for the magnetic field energy of interaction between *entire closed* current loops is applicable in the determination of the interaction energy between *different* current elements of the *same* current loop, which is the *finite* portion of the *self energy* of *the* current loop and will hereafter not be referred to as the self energy. Clearly, this is the only sensible physical generalization if one adopts as a physical principle the point of view that an element of current $\mathbf{w}^P$ at a point P cannot distinguish the origin of any portion of the magnetic induction vector $\mathbf{B}$ or its vector potential $\mathbf{A}$ at P, as we have already done in Sec. 6.3. In other words we have already assumed that the current element $\mathbf{w}^P$ cannot tell whether any portion of $\mathbf{A}$ comes from current elements $\mathbf{w}^Q$ associated with *its* closed loop or *any other* closed loop, in order to develop the magnetostatic field theory. Hence, this is not a further generalization of the description, but only a generalization of the expression for the energy that is consistent with the earlier generalization of the description introduced in Chapter 6 to obtain the field theory. In accordance with the foregoing the generalized expression for the total magnetostatic field energy is defined as

$$\mathcal{U} = \frac{1}{2}\sum_{n=1}^{N} \oint_{C^{(n)}} \mathbf{w}^{(n)} \cdot \mathbf{A}^{(n)} + \frac{1}{2}\sum_{n=1}^{N} \oint_{\substack{C^{(n)} \\ r^{(nn)} \neq 0}} \mathbf{w}^{(n)} \cdot \mathring{\mathbf{A}}^{(n)}, \tag{9.1.7}$$

where

$$\mathring{\mathbf{A}}^{(n)} = \oint_{\substack{C^{(n)} \\ r^{(nn)} \neq 0}} \frac{\mathbf{w}^{(n)}}{r^{(nn)}}. \tag{9.1.8}$$

As in the development of the field theory in Sec. 6.3 we now discretize[24] the loop integrals around the $C^{(n)}$ by writing the expression for $\mathcal{U}$ in (9.1.7) as a sum over T *different* current elements, $Y^{(n)}$ of which belong to the *same* closed current loop. In this way we obtain the expression

$$\mathcal{U} = \frac{1}{2}\sum_{t=1}^{T} \mathbf{w}^{(t)} \cdot \mathbf{A}^{(t)}, \tag{9.1.9}$$

where

$$A^{(t)} = \sum_{\substack{z=1 \\ z \neq t}} \frac{\mathbf{w}^{(z)}}{r^{zt}}, \tag{9.1.10}$$

and

$$T = \sum_{n=1}^{N} Y^{(n)}. \tag{9.1.11}$$

As we have before, we now proceed from the discrete model to a well behaved three-dimensional continuous distribution of current density **J** throughout all of space by means of the usual argument, which permits us to replace (9.1.9) by

$$\mathcal{U} = \frac{1}{2}\int_V \mathbf{J}\cdot\mathbf{A}\, dV. \tag{9.1.12}$$

In this expression the volume V either encompasses the current density in all space or goes to infinity, with the fields vanishing sufficiently rapidly for the infinite integral to converge, and no magnetization exists. Substituting from (7.3.12) into (9.1.12), we obtain

$$\mathcal{U} = \frac{1}{8\pi}\int_V (\nabla\times\mathbf{H})\cdot\mathbf{A}\, dV, \tag{9.1.13}$$

which, in indicial notation may be written

$$\mathcal{U} = \frac{1}{8\pi}\int_V e_{ijk}H_{k,j}A_i\, dV = \frac{1}{8\pi}\int_V e_{ijk}[(H_kA_i)_{,j} - H_kA_{i,j}]\, dV. \tag{9.1.14}$$

Since **A** and $\mathbf{B} = \mathbf{H}(\mathbf{M} = 0$ here) are continuous everywhere and $\nabla\mathbf{H}$ is continuous in the interior of the region if **J** has continuous first derivatives, which is all we are considering in this simple special case,[25] we may employ the divergence theorem to obtain

$$\mathcal{U} = \frac{1}{8\pi}\int_S e_{jki}n_jH_kA_i\, dS + \frac{1}{8\pi}\int_V H_ke_{kji}A_{i,j}\, dV, \tag{9.1.15}$$

which in vectorial notation takes the form

$$\mathcal{U} = \frac{1}{8\pi}\int_S \mathbf{n}\cdot\mathbf{H}\times\mathbf{A}\, dS + \frac{1}{8\pi}\int_V \mathbf{H}\cdot(\nabla\times\mathbf{A})\, dV. \tag{9.1.16}$$

Now we let the surface S recede to infinity and note that the fields **A** and **H** drop off as $1/r_1$ and $1/r_1^2$, respectively, so that the product drops off as $1/r_1^3$ while the area goes up as r_1^2. Consequently, in the limit the surface integral vanishes, and we are left with

$$\mathcal{U} = \frac{1}{8\pi}\int_V \mathbf{H}\cdot\mathbf{B}\, dV, \tag{9.1.17}$$

in Gaussian units. Since in the situation we are now considering there is no **M**,

$$\mathbf{H} = \mathbf{B}, \tag{9.1.18}$$

in Gaussian units. Hence, we have

$$\mathcal{U} = \frac{1}{8\pi}\int_V \mathbf{B}\cdot\mathbf{B}\, dV = \frac{1}{8\pi}\int_V \mathbf{H}\cdot\mathbf{H}\, dV, \tag{9.1.19}$$

which shows that in this case[26] ($\mathbf{M} = 0$), the magnetic field energy due to a current distribution somewhere in space is mathematically equivalent to a volume magnetic energy density in all of space, even vacuum, of magnitude

$$\mathbf{B}\cdot\mathbf{B}/8\pi, \tag{9.1.20}$$

in Gaussian units.

Now we proceed just as in the electrostatic case, and postulate that this mathematical equivalence is more than an equivalence but is a physical fact. This means that we associate with each volume element of free space (vacuum) and/or volume element of current density $\mathbf{J}$ in which a magnetic induction field $\mathbf{B}$ exists, a magnetic field energy density of $B^2/8\pi$.

9.2 Energy Resulting from Distributions of Current and Magnetization

We have so far said nothing about the magnetic field energy density in the presence of magnetic dipole moments (or magnetization). As in the case of electrostatics we first consider the energy of interaction of *a* magnetic dipole moment in a magnetic induction field. To this end we consider the usual diagram shown in Figure 25 and use the expression for magnetostatic energy of a current loop in the form

$$\mathscr{U} = \oint_C I\, d\mathbf{s}\cdot\mathbf{A} = I \oint_C \mathbf{A}\cdot d\mathbf{s}, \tag{9.2.1}$$

to which we apply Stokes theorem and obtain

$$\mathscr{U} = I \int_{S_0} \mathbf{n}\cdot\nabla \times \mathbf{A}\, dS. \tag{9.2.2}$$

We now take the limit of (9.2.2) as $S_0 \to 0$, $I \to \infty$ with $\mathbf{n}$ fixed and, with the aid of the law of the mean or Taylor's theorem, find

$$\mathscr{U} = \lim_{\substack{S_0 \to 0 \\ I \to \infty}} I\mathbf{n}\cdot S_0 \nabla \times \mathbf{A} = \mathbf{m}\cdot\mathbf{B}, \tag{9.2.3}$$

as the expression for the energy of a magnetic dipole moment $\mathbf{m}$ in an *external* magnetic induction field $\mathbf{B}$. Thus it is clear that the energy of interaction of magnetization with the magnetic induction field is a consequence of orientation as opposed to position, as we expected from our previous experience in the electrostatic case.

In order to find the expression for the magnetic field energy when the magnetic induction field is *internal* to the system of current and magnetization we employ the generalization of our general formula (9.1.5) for a system of current loops, which we write in the form

$$\begin{aligned}\mathcal{U} = \tfrac{1}{2} I^{(1)} \oint_{C^{(1)}} d\mathbf{s}^{(1)} \cdot \left[\oint_{C^{(2)}} \frac{d\mathbf{s}^{(2)}}{r^{12}} + I^{(3)} \oint_{C^{(3)}} \frac{d\mathbf{s}^{(3)}}{r^{13}} + \cdots + I^{(N)} \oint_{C^{(N)}} \frac{d\mathbf{s}^{(N)}}{r^{1N}} \right] \\ + \tfrac{1}{2} I^{(2)} \oint_{C^{(2)}} d\mathbf{s}^{(2)} \cdot \left[I^{(1)} \oint_{C^{(1)}} \frac{d\mathbf{s}^{(1)}}{r^{21}} + I^{(3)} \oint_{C^{(3)}} \frac{d\mathbf{s}^{(3)}}{r^{23}} + \cdots + I^{(N)} \oint_{C^{(N)}} \frac{d\mathbf{s}^{(N)}}{r^{2N}} \right] + \cdots \\ + \tfrac{1}{2} I^{(N)} \oint_{C^{(N)}} d\mathbf{s}^{(N)} \cdot \left[I^{(1)} \oint_{C^{(1)}} \frac{d\mathbf{s}^{(1)}}{r^{N1}} + I^{(2)} \oint_{C^{(2)}} \frac{d\mathbf{s}^{(2)}}{r^{N2}} \right. \\ \left. + \cdots + I^{(N-1)} \oint_{C^{(N-1)}} \frac{d\mathbf{s}^{(N-1)}}{r^{N(N-1)}} \right] + \sum_{n=1}^{N} u^{S(n)}, \end{aligned} \tag{9.2.4}$$

where $u^{S(n)}$ is the finite portion of the self-energy of each *extended* current loop, which we have already discussed. This term has no influence on the expression for the energy contribution due to point dipoles because, as we already know, the extension of the current loop which forms the point dipole vanishes. We now let the second current loop become a dipole in the usual way, i.e., we consider the diagram shown in Figure 25 and apply the integral theorem in (7.1.2) to each term in (9.2.4) of the form

$$I^{(2)} \oint_{C^{(2)}} \frac{d\mathbf{s}^{(2)}}{r^{n2}}, \tag{9.2.5}$$

and take the limit as $S_0^{(2)} \to 0$, $I^{(2)} \to \infty$ with $\mathbf{n}^{(2)}$ fixed in such a way that

$$\lim_{\substack{S_0^{(2)} \to 0 \\ I^{(2)} \to \infty}} \mathbf{n}^{(2)} I^{(2)} S_0^{(2)} = \mathbf{n}^{(2)} m^{(2)} = \mathbf{m}^{(2)}. \tag{9.2.6}$$

Under the foregoing circumstances from (7.1.2) we have

$$I^{(2)} \oint_{C^{(2)}} \frac{d\mathbf{s}^{(2)}}{r^{n2}} = I^{(2)} \oint_{S_0^{(2)}} \mathbf{n}^{(2)} \times \nabla^{(2)} \left(\frac{1}{r^{n2}} \right) dS, \tag{9.2.7}$$

and in the limit we find

$$\begin{aligned}\lim_{\substack{S_0^{(2)} \to 0 \\ I^{(2)} \to \infty \\ \mathbf{n}^{(2)}\ \text{fixed}}} I^{(2)} \oint_{C^{(2)}} \frac{d\mathbf{s}^{(2)}}{r^{n2}} &= \lim_{\substack{S_0^{(2)} \to 0 \\ I^{(2)} \to \infty}} I^{(2)} \mathbf{n}^{(2)} S_0^{(2)} \times \nabla^{(2)} \left(\frac{1}{r^{n2}} \right) \\ &= \mathbf{m}^{(2)} \times \nabla^{(2)} \left(\frac{1}{r^{n2}} \right). \end{aligned} \tag{9.2.8}$$

Before substituting back into the energy expression we observe that in the sum of (9.2.4) $\mathbf{m}^{(2)}$ occurs in two different ways, i.e., either from

$$\tfrac{1}{2} I^{(1)} \oint_{C^{(1)}} d\mathbf{s}^{(1)} \cdot I^{(2)} \oint_{C^{(2)}} \frac{d\mathbf{s}^{(2)}}{r^{12}}, \tag{9.2.9}$$

or from

$$\tfrac{1}{2}I^{(2)}\oint_{C^{(2)}}\frac{d\mathbf{s}^{(2)}}{r^{21}}\cdot I^{(1)}\oint_{C^{(1)}}d\mathbf{s}^{(1)}. \tag{9.2.10}$$

In the first case in the limit we have

$$\tfrac{1}{2}I^{(1)}\oint_{C^{(1)}}d\mathbf{s}^{(1)}\cdot\left[\mathbf{m}^{(2)}\times\nabla^{(2)}\left(\frac{1}{r^{12}}\right)\right]=\tfrac{1}{2}I^{(1)}\oint_{C^{(1)}}d\mathbf{s}^{(1)}\cdot\left[\nabla^{(1)}\left(\frac{1}{r^{12}}\right)\times\mathbf{m}^{(2)}\right], \tag{9.2.11}$$

in which the quantity in brackets is the expression for **A** at circuit loop 1 due to a dipole moment $\mathbf{m}^{(2)}$ at point 2. In the second case in the limit we have

$$\oint_{C^{(1)}}\frac{1}{2}\left[\mathbf{m}^{(2)}\times\nabla^{(2)}\left(\frac{1}{r^{21}}\right)\right]\cdot I^{(1)}d\mathbf{s}^{(1)}, \tag{9.2.12}$$

which clearly may be written in the form

$$\oint_{C^{(1)}}\frac{1}{2}m_j^{(2)}e_{jki}\frac{\partial}{\partial x_k^{(2)}}I^{(1)}\frac{ds_i^{(1)}}{r^{21}}=\tfrac{1}{2}\mathbf{m}^{(2)}\cdot\nabla^{(2)}\times I^{(1)}\oint_{C^{(1)}}\frac{d\mathbf{s}^{(1)}}{r^{21}}, \tag{9.2.13}$$

and the term after the cross product is clearly the value for **A** at point 2 due to the circuit loop 1. From the foregoing it is clear that if we let the third circuit loop become a dipole moment in the limit we would have

$$\lim_{\substack{S_0^{(3)}\to 0\\ I^{(3)}\to\infty\\ n^{(3)}\text{ fixed}}} I^{(3)}\oint_{C^{(3)}}\frac{d\mathbf{s}^{(3)}}{r^{n3}}=\mathbf{m}^{(3)}\times\nabla^{(3)}\left(\frac{1}{r^{n3}}\right). \tag{9.2.14}$$

Under these circumstances there would be terms in the second (and third) line of our general expression in (9.2.4) of the form

$$\tfrac{1}{2}I^{(2)}\oint_{C^{(2)}}d\mathbf{s}^{(2)}\cdot I^{(3)}\oint_{C^{(3)}}\frac{d\mathbf{s}^{(3)}}{r^{23}}. \tag{9.2.15}$$

From the first limit in (9.2.13) we have

$$\tfrac{1}{2}\mathbf{m}^{(2)}\cdot\nabla^{(2)}\times I^{(3)}\oint_{C^{(3)}}\frac{d\mathbf{s}^{(3)}}{r^{23}}, \tag{9.2.16}$$

and from the second limit in (9.2.14) we have

$$\tfrac{1}{2}\mathbf{m}^{(2)}\cdot\nabla^{(2)}\times\left[\mathbf{m}^{(3)}\times\nabla^{(3)}\left(\frac{1}{r^{23}}\right)\right]=\tfrac{1}{2}\mathbf{m}^{(2)}\cdot\nabla^{(2)}\times\nabla^{(2)}\times\frac{\mathbf{m}^{(3)}}{r^{23}}, \tag{9.2.17}$$

and, as we know, the term $\nabla^{(2)}\times(\mathbf{m}^{(3)}/r^{23})$ is the magnetic vector potential **A** at point 2 due to a dipole moment at point 3. This is the expression that must go in the second line of our general sum in (9.2.4). Clearly the expression that must go in the third line may be obtained from (9.2.17) simply by interchanging the roles of 2 and 3, thus

$$\tfrac{1}{2}\mathbf{m}^{(3)}\cdot\nabla^{(3)} \times \nabla^{(3)} \times \frac{\mathbf{m}^{(2)}}{r^{32}}. \tag{9.2.18}$$

Incorporating all of the foregoing results for dipoles in our general sum in (9.2.4) we obtain

$$\begin{aligned}
\mathscr{U} = {} & \tfrac{1}{2}I^{(1)}\oint_{C^{(1)}} d\mathbf{s}^{(1)}\cdot\Bigg[\nabla^{(1)} \times \frac{\mathbf{m}^{(2)}}{r^{12}} + \nabla^{(1)} \times \frac{\mathbf{m}^{(3)}}{r^{13}} \\
& + I^{(4)}\oint_{C^{(4)}} \frac{d\mathbf{s}^{(4)}}{r^{14}} + \cdots + I^{(N)}\oint_{C^{(N)}} \frac{d\mathbf{s}^{(N)}}{r^{1N}}\Bigg] \\
& + \tfrac{1}{2}\mathbf{m}^{(2)}\cdot\nabla^{(2)} \times \Bigg[I^{(1)}\oint_{C^{(1)}} \frac{d\mathbf{s}^{(1)}}{r^{21}} + \nabla^{(2)} \times \frac{\mathbf{m}^{(3)}}{r^{23}} \\
& + I^{(4)}\oint_{C^{(4)}} \frac{d\mathbf{s}^{(4)}}{r^{24}} + \cdots + I^{(N)}\oint_{C^{(N)}} \frac{d\mathbf{s}^{(N)}}{r^{2N}}\Bigg] \\
& + \tfrac{1}{2}\mathbf{m}^{(3)}\cdot\nabla^{(3)} \times \Bigg[I^{(1)}\oint_{C^{(1)}} \frac{d\mathbf{s}^{(1)}}{r^{31}} + \nabla^{(3)} \times \frac{\mathbf{m}^{(2)}}{r^{32}} \\
& + I^{(4)}\oint_{C^{(4)}} \frac{d\mathbf{s}^{(4)}}{r^{34}} + \cdots + I^{(N)}\oint_{C^{(N)}} \frac{d\mathbf{s}^{(N)}}{r^{3N}}\Bigg] + \cdots \\
& + \tfrac{1}{2}I^{(N)}\oint_{C^{(N)}} d\mathbf{s}^{(N)}\cdot\Bigg[I^{(1)}\oint_{C^{(1)}} \frac{d\mathbf{s}^{(1)}}{r^{N1}} + \nabla^{(N)} \times \frac{\mathbf{m}^{(2)}}{r^{N2}} \\
& + \nabla^{(N)} \times \frac{\mathbf{m}^{(3)}}{r^{N3}} + I^{(4)}\oint_{C^{(4)}} \frac{d\mathbf{s}^{(4)}}{r^{N4}} + \cdots + I^{(N-1)}\oint_{C^{(N-1)}} \frac{d\mathbf{s}^{(N-1)}}{r^{N(N-1)}}\Bigg] \\
& + \sum_n u^{S(n)}.
\end{aligned} \tag{9.2.19}$$

An examination of this expression and previous expressions (6.2.19) and (7.1.8) for **A** for systems of current loops and point dipoles reveals that the expressions in brackets are the magnetic potential **A** at the current loop or dipole due to *all other* current loops and dipoles. Hence it is clear that for a system of N current loops and D point dipoles we may write

$$\mathscr{U} = \frac{1}{2}\sum_{n=1}^{N} I^{(n)}\oint_{C^{(n)}} d\mathbf{s}^{(n)}\cdot\mathbf{A}^{(n)} + \frac{1}{2}\sum_{d=1}^{D} \mathbf{m}^{(d)}\cdot\mathbf{B}^{(d)} + \sum_{n=1}^{N} u^{S(n)}, \tag{9.2.20}$$

since the expression for the finite portion of the self energy of the extended current loops does not affect the field at, nor is it affected by, the point dipoles. Clearly then, in the same way that we did before and for the same reason we may combine the first and last sums in (9.2.20) and write

$$\mathscr{U} = \frac{1}{2}\sum_{t=1}^{T} \mathbf{w}^{(t)}\cdot\mathbf{A}^{(t)} + \frac{1}{2}\sum_{d=1}^{D} \mathbf{m}^{(d)}\cdot\mathbf{B}^{(d)}. \tag{9.2.21}$$

When we proceed from the discrete current loop and point dipole point of view to the continuous current density and dipole moment density (or field) point of view, we must remember that the magnetic induction $\mathbf{B}$ in the magnetized region does not determine the force on a current element because the local induction field

$$\mathbf{B}^L = -\int_\Sigma \nabla^P\left(\frac{1}{r_1}\right) \times (\mathbf{n} \times \mathbf{M})\,dS, \tag{9.2.22}$$

which depends on the shape of the surface Σ surrounding the vanishingly small element of volume, has been excluded from the vector $\mathbf{B}$ appearing in the field equations. As a consequence of this there must be a local energy density depending on $\mathbf{M}$ at each point of the magnetized continuum. In addition to this local energy, which must be quadratic in $\mathbf{M}$, there is another very important contribution to the local energy density of the magnetized matter, which is due to the creation of the magnetization at each point of the matter by the magnetic induction field and has not been included in anything we have done heretofore. This latter energy is the energy of induced magnetization, which is a result of the microscopic reorientation of electron spin by $\mathbf{B}$. Let us represent the combination of both local matter energies by the expression u^L.

Now, let us determine the expression for the energy in the general case of continuous distributions of current density $\mathbf{J}$ and magnetization $\mathbf{M}$. To this end we consider a collection of regions of continuous distributions of current density $\mathbf{J}$ and magnetization $\mathbf{M}$. When we apply Eq. (9.2.21) for magnetostatic field energy to the situation under consideration and employ the argument to go from the discrete model to the continuous distributions, we obtain

$$\mathscr{U} = \frac{1}{2}\int_{\mathscr{V}} \mathbf{J}\cdot\mathbf{A}\,dV + \frac{1}{2}\int_{\mathscr{V}} \mathbf{M}\cdot\mathbf{B}\,dV + \int_{\mathscr{V}} u^L(\mathbf{M})\,dV, \tag{9.2.23}$$

where $\mathscr{V}$ represents the collection of all volume distributions of continuous current density $\mathbf{J}$ and magnetization $\mathbf{M}$ and u^L accounts for the aforementioned local magnetic energy of the matter. The boundary conditions between any regions of continuous distributions of $\mathbf{J}$ and/or $\nabla \times \mathbf{M}$ and in which their first spatial derivatives are continuous are given by Eqs. (7.5.4) and (7.5.5), which enable us to write

$$\mathbf{n}\cdot(\mathbf{B}^+ - \mathbf{B}^-) = 0, \tag{9.2.24}$$

$$\mathbf{n} \times (\mathbf{H}^+ - \mathbf{H}^-) = 0, \tag{9.2.25}$$

for the surfaces of each such region. In the interior of such regions Eqs. (7.3.12)–(7.3.14) hold. Before proceeding we also note from Sec. 7.2 and Eq. (7.5.11) that **A** is continuous throughout the entire space. Now substituting from (7.3.12) for **J** and introducing indicial notation, we obtain

$$\mathscr{U} = \frac{1}{8\pi}\int_{\mathscr{V}} e_{ijk} H_{k,j} A_i dV + \frac{1}{2}\int_{\mathscr{V}} M_k B_k dV + \int_{\mathscr{V}} u^L(\mathbf{M}) dV. \tag{9.2.26}$$

The first volume integral in (9.2.26) may be written

$$\int_{\mathscr{V}} e_{ijk}[(H_k A_i)_{,j} - H_k A_{i,j}] dV, \tag{9.2.27}$$

for each region V in which both **J** and $\nabla \times \mathbf{M}$ and their first spatial derivatives are continuous. Application of the divergence theorem to (9.2.27) yields

$$\sum_{n=1}^{R} \int_{S^{(n)}} e_{ijk} n_j H_k A_i dS + \int_{\mathscr{S}} e_{ijk} n_j H_k A_i dS - \int_{\mathscr{S}} e_{ijk} H_k A_{i,j} dV, \tag{9.2.28}$$

where $S^{(n)}$ represents the surface surrounding the nth volume in which the aforementioned quantities are continuous and n_j represents the components of the *outwardly* directed unit normal to the surface of the nth region and $\mathscr{S}$ represents the entire outer surface which will recede to infinity. Clearly, any portion of an $S^{(n)}$ separates two different regions of continuity, say the nth from the $(n+1)$th. Then the outward normal to one volume at any point of an $S^{(n)}$ is the inward normal to the other. Hence, we may write

$$\sum_{n=1}^{R} \int_{S^{(n)}} e_{ijk} n_j H_k A_i dS = \sum_{m=1}^{W} \int_{S^{(m)}} e_{ijk} n_j^+ (H_k^+ A_i^+ - H_k^- A_i^-) dS, \tag{9.2.29}$$

where n_j^+ represents the unit normal directed from the $-$ to the $+$ side of any surface $S^{(m)}$. Since A_i is continuous, we have

$$\int_{S^{(m)}} e_{ijk} n_j^+ (H_k^+ A_i^+ - H_k^- A_i^-) dS = \int_{S^{(m)}} A_i e_{ijk} n_j^+ (H_k^+ - H_k^-) dS, \tag{9.2.30}$$

across each surface and by virtue of (9.2.25), we find

$$\int_{S^{(m)}} e_{ijk} n_j^+ (H_k^+ A_i^+ - H_k^- A_i^-) dS = 0. \tag{9.2.31}$$

Moreover, as $\mathscr{S}$ recedes to infinity, **A** and **H** drop off as $1/r_1$ and $1/r_1^2$, respectively, so that the product drops off as $1/r_1^3$ while the area of $\mathscr{S}$ goes up as r_1^2, so that we have

$$\lim_{r_1 \to \infty} \int_{\mathscr{S}} e_{ijk} n_j H_k A_i dS \to 0. \tag{9.2.32}$$

Collecting these results and substituting back in the energy expression in (9.2.26), we find

$$\mathscr{U} = \int_{\mathscr{V}} \left[\frac{1}{8\pi} e_{kji} H_k A_{i,j} + \tfrac{1}{2} M_k B_k + u^L(\mathbf{M}) \right] dV, \tag{9.2.33}$$

which, with (7.2.14) may be written in the vectorial form

$$\mathscr{U} = \int_{\mathscr{V}} \left[\frac{1}{8\pi} \mathbf{H} \cdot \mathbf{B} + \tfrac{1}{2} \mathbf{M} \cdot \mathbf{B} + u^L(\mathbf{M}) \right] dV, \tag{9.2.34}$$

and with the aid of (7.3.13), we obtain

$$\mathscr{U} = \int_{\mathscr{V}} \left[\frac{\mathbf{B} \cdot \mathbf{B}}{8\pi} + u^L(\mathbf{M}) \right] dV. \tag{9.2.35}$$

This last expression in (9.2.35) shows that in the most general case, i.e., when **J** and **M** are present, the stored energy in the field and the matter is just the ordinary magnetostatic energy of free-space $B^2/8\pi$ and the local material energy of induced magnetization plus the energy associated with the local induction field $\mathbf{B}^L$ that has been left out in defining **B**.

9.3 Energy in Linear Magnetostatics

The expression in (9.2.35) we have obtained is perfectly general and is valid in the nonlinear case. However, it should be noted that in linear magnetostatics we have

$$u^L = -\frac{1}{2\zeta} M_i M_i, \tag{9.3.1}$$

$$B_i = \frac{1}{\zeta} M_i, \tag{9.3.2}$$

in the isotropic case and

$$u^L = -\tfrac{1}{2} \gamma_{ij} M_i M_j, \tag{9.3.3}$$

$$B_i = \gamma_{ij} M_j, \tag{9.3.4}$$

in the anisotropic case, where γ_{ij} is the reciprocal of ζ_{ij}. In either case we have

$$u^L = -\tfrac{1}{2} B_i M_i, \tag{9.3.5}$$

which with (9.2.35) enables us to write

$$\mathscr{U} = \int_V \left[\frac{B_k B_k}{8\pi} - \frac{1}{2} B_k M_k \right] dV = \frac{1}{8\pi} \int_V B_k H_k \, dV, \tag{9.3.6}$$

which is the well-known expression for the energy in *linear* magnetostatics.

Clearly, by comparison with the susceptibility and permeability tensors ψ_{ij} and μ_{ij}, from (9.3.4) we have

$$\psi_{ij} = \zeta_{ik}\mu_{kj}, \tag{9.3.7}$$

$$\mu_{ij} = \gamma_{ik}\psi_{kj}. \tag{9.3.8}$$

Hence ζ_{ij} and, of course, γ_{ij} can be determined if ψ_{ij} and, hence, μ_{ij} are known. Moreover, as a consequence of the existence of the stored energy function γ_{ij} (and, of course, ζ_{ij}) and μ_{ij} (and, of course, ψ_{ij}) are symmetric, i.e.,

$$\gamma_{ij} = \gamma_{ji}, \quad \zeta_{ij} = \zeta_{ji}, \quad \mu_{ij} = \mu_{ji}, \quad \psi_{ij} = \psi_{ji}, \tag{9.3.9}$$

for the same reason as in electrostatics. This treatment holds, of course, for linear paramagnetic and diamagnetic materials only. This ends our development of magnetostatics and brings us to electromagnetics.

III

ELECTROMAGNETICS

CHAPTER 10

The Electromagnetic Field Equations

10.1 Time Dependence

In electrostatics we have Eq. (3.5.3), which for convenience we rewrite here as

$$\oint_C \mathbf{E}\cdot d\mathbf{r} = 0, \tag{10.1.1}$$

where C is *any* closed loop. It is also known that electrostatic fields of this nature, i.e., satisfying (10.1.1), cannot be associated with steady (non-time-dependent) currents **J** because flowing currents are associated with the dissipation of energy and electrostatic fields cannot provide this energy. This is not surprising since **E** was defined for static conditions only, i.e., when the charge density ρ had no velocity **v** and, hence, no current **J** was flowing. We also observe from our work in magnetostatics that a steady current **J** flowing in a wire generates a magnetic induction field **B** by virtue of Eq. (7.3.12); but a magnetic induction field **B** does not generate a current **J**. However, Faraday discovered that when the magnetic induction field **B** varied in time, a current was induced in a wire. Moreover, he found a precise quantitative statement describing the effect. However, the form stated by Faraday was for a particular experimental system and, consequently, is not as useful analytically as the form in which it was stated by Maxwell.

To state Faraday's law in the form given it by Maxwell we consider the current loop (circuit) shown in Figure 30 and define the electromotive force (emf) by

$$\mathscr{E}^e = \oint_C \mathbf{E}\cdot d\mathbf{r}, \tag{10.1.2}$$

where this expression is in esu since **E** is in esu. In emu we have

$$\mathscr{E}^M = c\oint_C \mathbf{E}\cdot d\mathbf{r}, \tag{10.1.3}$$

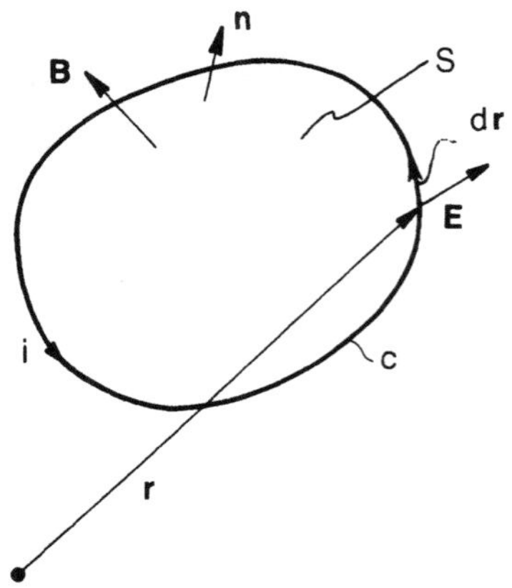

FIGURE 30. Closed circuit for the statement of Faraday's Law.

since $q^M = q^E/c$ and

$$\mathbf{E}^E = \frac{\mathbf{F}}{q^E}, \quad \mathbf{E}^M = \frac{\mathbf{F}}{q^M} = c\frac{\mathbf{F}}{q^E} = c\mathbf{E}^E. \tag{10.1.4}$$

We now define the magnetic flux Φ^M threading the loop C by

$$\Phi^M = \int_S \mathbf{n}\cdot\mathbf{B}\,dS, \tag{10.1.5}$$

where S is any surface through the current loop C and $\mathbf{n}$ is positive for a counterclockwise circuit around the loop. The quantity Φ^M is called the magnetic flux of the circuit. Then Maxwell' statement of Faraday's law is

$$\mathscr{E}^M = -\,\partial\Phi^M/\partial t, \tag{10.1.6}$$

and substituting from (10.1.3) and (10.1.5) into (10.1.16), we obtain

$$c\oint_C \mathbf{E}\cdot d\mathbf{r} = -\frac{\partial}{\partial t}\int_S \mathbf{n}\cdot\mathbf{B}\,dS. \tag{10.1.7}$$

We now postulate that this equation, written for a current loop C, is applicable to *any* arbitrary circuit C enclosing an area S in space. Application of Stokes' theorem to (10.1.7) yields

$$c\int_S \mathbf{n}\cdot\nabla\times\mathbf{E}\,dS = -\frac{\partial}{\partial t}\int_S \mathbf{n}\cdot\mathbf{B}\,dS, \tag{10.1.8}$$

and since S is stationary we have

$$\int_S \mathbf{n}\cdot\left[c\nabla\times\mathbf{E} + \frac{\partial\mathbf{B}}{\partial t}\right]dS = 0, \tag{10.1.9}$$

which, since $\mathbf{n}$ and S are arbitrary, yields

$$\nabla \times \mathbf{E} = -\frac{1}{c}\frac{\partial \mathbf{B}}{\partial t}. \tag{10.1.10}$$

Equation (10.1.10) is one of Maxwell's vector equations. Clearly, it is a direct consequence of Faraday's law.

In magnetostatics we had steady currents so that there could be no charge accumulation with time in a volume element. Under these circumstances from (6.3.22) we had

$$\int_S \mathbf{n}\cdot\mathbf{J}\,dS = \int_V \nabla\cdot\mathbf{J}\,dV = 0, \tag{10.1.11}$$

where $\mathbf{J}$ is current in magnetic units, which is defined by Eq. (6.3.8) and is essentially given by

$$\mathbf{J}\,dV = \sum_n q^{M(n)}\mathbf{v}^{(n)}. \tag{10.1.12}$$

Hence the $\mathbf{J}$ we have been using in Chapters 6–9 should more properly have been denoted $\mathbf{J}^M$, i.e.,

$$\mathbf{J} \equiv \mathbf{J}^M, \tag{10.1.13}$$

since in magnetostatics we have been using emu. In esu we have $\mathbf{J}^E$, which is defined by

$$\mathbf{J}^E dV = \sum_n q^{E(n)}\mathbf{v}^{(n)}, \tag{10.1.14}$$

so that we may write

$$\mathbf{J}^E dV = \sum_n q^{E(n)}\mathbf{v}^{(n)} = c\sum_n q^{M(n)}\mathbf{v}^{(n)} = c\mathbf{J}^M dV. \tag{10.1.15}$$

Hence, we have

$$\mathbf{J}^E = c\mathbf{J}^M. \tag{10.1.16}$$

From now on we will write current densities in esu, whereas heretofore we have been writing current densities in emu. Thus for steady currents in emu, we have

$$\int_S \mathbf{n}\cdot\mathbf{J}^M dS = \int_V \nabla\cdot\mathbf{J}^M dV = 0, \tag{10.1.17}$$

and in esu we have

$$c\int_S \mathbf{n}\cdot\mathbf{J}^E dS = c\int_V \nabla\cdot\mathbf{J}^E dV = 0. \tag{10.1.18}$$

Note that the magnetostatic field equation (7.3.12), i.e.,

$$\nabla \times \mathbf{H} = 4\pi \mathbf{J}, \tag{10.1.19}$$

has been written in emu and in our new notation should be written

$$\nabla \times \mathbf{H} = 4\pi \mathbf{J}^M. \tag{10.1.20}$$

Now, if we substitute from (10.1.16) for the current density in esu, we obtain

$$\nabla \times \mathbf{H} = \frac{4\pi}{c} \mathbf{J}^E, \tag{10.1.21}$$

as the field equation of magnetostatics.

Now, it is clear that if we do not have steady currents, charge can accumulate in time in a volume element. Consider the region shown in Figure 31 where $\rho \equiv \rho^E$ is the charge density in esu. Clearly the total charge Q in V is given by

$$Q = \int_V \rho \, dV, \tag{10.1.22}$$

and the rate of *increase* of charge $\partial Q/\partial t$ is given by

$$\frac{\partial Q}{\partial t} = \int_V \frac{\partial \rho}{\partial t} dV, \tag{10.1.23}$$

which equation is in esu. Now, $\mathbf{v}$ is the velocity at which charge density ρ is flowing at the generic point P. Clearly then

$$\rho \mathbf{v} \equiv \mathbf{J}^E. \tag{10.1.24}$$

The conservation of charge for a *stationary* volume V says that the rate at which the total charge is *decreased* in V is equal to the rate at which charge flows out of that stationary volume V. Consequently we may write

$$-\frac{\partial Q}{\partial t} = \int_S \mathbf{n} \cdot \rho \mathbf{v} \, dS, \tag{10.1.25}$$

which with (10.1.23) and (10.1.24) enables us to write

$$-\int_V \frac{\partial \rho}{\partial t} dV = \int_S \mathbf{n} \cdot \mathbf{J}^E dS, \tag{10.1.26}$$

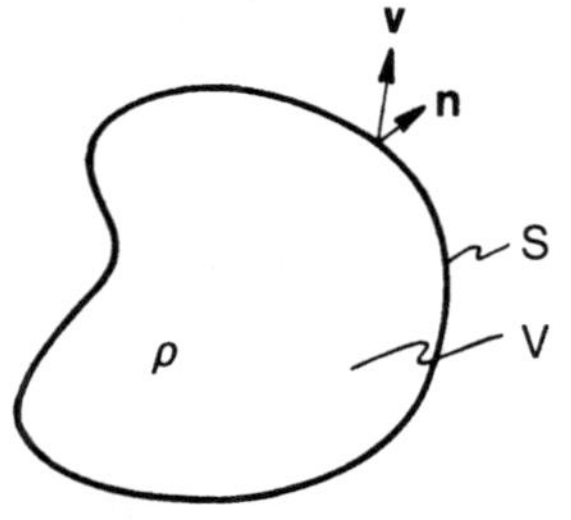

FIGURE 31. Arbitrary region containing time-dependent moving charge.

and an application of the divergence theorem yields

$$\int_V \left(\frac{\partial \rho}{\partial t} + \nabla \cdot \mathbf{J}^E\right) dV = 0. \tag{10.1.27}$$

Since this equation holds for any arbitrary volume V, we have

$$\nabla \cdot \mathbf{J}^E + \partial \rho / \partial t = 0, \tag{10.1.28}$$

at each point. Equation (10.1.28) is the generalization of $\nabla \cdot \mathbf{J}^E = 0$ for steady currents to time-dependent currents and charge.

Now we are interested in the time-dependent generalization of the magnetostatic field equation (10.1.21). Note that the divergence of (10.1.21) yields

$$\nabla \cdot \mathbf{J}^E = 0, \tag{10.1.29}$$

which is inconsistent with Eq. (10.1.28) for nonsteady currents. At this point Maxwell comes on the scene and makes a crucial contribution. His first step in resolving this dilemma was to note that in electrostatics the field equation (3.3.8), i.e.,

$$\nabla \cdot \mathbf{D} = 4\pi\rho, \tag{10.1.30}$$

exists, which relates the electric charge density to the electric displacement. Maxwell then *assumed* that Eq. (10.1.30) remains valid in the time-dependent case. Clearly then, this means that there must be another vector field in the magnetostatic field equation (10.1.21), the existence of which had not been revealed in any experiment up to that time. Let this vector field be G. We then rewrite Eq. (10.1.21) in the form which we expect to be valid in the time-dependent (or dynamic) case, i.e.,

$$\nabla \times \mathbf{H} = \frac{4\pi}{c}\mathbf{J}^E + \mathbf{G}. \tag{10.1.31}$$

Taking the divergence of (10.1.31), we obtain

$$\frac{4\pi}{c}\nabla \cdot \mathbf{J}^E + \nabla \cdot \mathbf{G} = 0, \tag{10.1.32}$$

and substituting the time derivative of (10.1.30) (which has been assumed to hold in the dynamic case) along with (10.1.28) into (10.1.32), we find

$$\nabla \cdot \left[-\frac{1}{c}\frac{\partial \mathbf{D}}{\partial t} + \mathbf{G} \right] = 0, \tag{10.1.33}$$

which enables us to write

$$\mathbf{G} = \frac{1}{c}\frac{\partial \mathbf{D}}{\partial t} + \nabla \times \mathbf{W}, \tag{10.1.34}$$

and **W** is taken to vanish since it is not needed for consistency. Thus, from (10.1.34) with $\mathbf{W}=0$ and (10.1.31) we now have the other Maxwell electromagnetic field equation

$$\nabla \times \mathbf{H} = \frac{1}{c}\left[4\pi \mathbf{J}^{E} + \frac{\partial \mathbf{D}}{\partial t}\right]. \tag{10.1.35}$$

We now have Maxwell's full electromagnetic field equations. They are the vector field equations (10.1.35) and (10.1.10), which because of their importance we rewrite here

$$\nabla \times \mathbf{H} = \frac{1}{c}\left[4\pi \mathbf{J} + \frac{\partial \mathbf{D}}{\partial t}\right], \tag{10.1.36}$$

$$\nabla \times \mathbf{E} = -\frac{1}{c}\frac{\partial \mathbf{B}}{\partial t}, \tag{10.1.37}$$

and the auxiliary scalar equations (10.1.30) and (7.3.14), which for completeness we also rewrite here

$$\nabla \cdot \mathbf{D} = 4\pi\rho, \quad \nabla \cdot \mathbf{B} = 0. \tag{10.1.38}$$

We now note that in electromagnetism the first of (10.1.38) simply serves to define the charge density ρ and the time derivative of the second is satisfied identically by virtue of (10.1.37). The vector fields **E**, **D**, **H** and **B** are, in general, related by Eqs. (3.3.9) and (7.3.13), which for completeness we rewrite here

$$\mathbf{D} = \mathbf{E} + 4\pi\mathbf{P}, \quad \mathbf{B} = \mathbf{H} + 4\pi\mathbf{M}, \tag{10.1.39}$$

where **P** and **M** are the previously defined polarization and magnetization, respectively, and belong to the *matter*, as opposed to the electric field **E** and magnetic induction field **B**, which belong to the *space*. These equations in this form are quite general but cannot be solved unless there are relations between **P** and **E**, **M** and **B** and **J** and **E**.

10.2 Linearized Electromagnetism

In the usual treatments of electromagnetism linear relations are assumed in each case. In the case of **P** and **E** and **M** and **B** the linear relations are exactly the same as in electrostatics and magnetostatics, which are given in Secs. 3.4 and 7.4, respectively. In the anisotropic case these relations take the respective forms

$$P_i = \chi_{ij}E_j, \quad D_i = \varepsilon_{ij}E_j, \tag{10.2.1}$$

$$M_i = \zeta_{ij}B_j, \quad M_i = \psi_{ij}H_j, \quad B_i = \mu_{ij}H_j. \tag{10.2.2}$$

In the case of **J** and **E**, the linear relation is

$$J_i = \sigma_{ij}E_j, \tag{10.2.3}$$

where σ_{ij} is the electrical conductivity tensor. The reciprocal of σ_{ij} is r_{ij} which is the electrical resistivity tensor. Equation (10.2.3) is the field equation of Ohm's law. If the material is isotropic or cubic

$$\sigma_{ij} = \sigma\delta_{ij}, \tag{10.2.4}$$

and σ is the electrical conductivity. The expression

$$E_k J_k = E_k \sigma_{kl} E_l, \tag{10.2.5}$$

is called the Joule heat term and represents the rate of dissipation of electromagnetic energy in this linear case.[27] Thus in the linear case, in addition to Maxwells equations (10.1.36)–(10.1.38), we have the linear electromagnetic constitutive relations

$$D_i = \varepsilon_{ij} E_j, \quad B_i = \mu_{ij} H_j, \quad J_i = \sigma_{ij} E_j, \tag{10.2.6}$$

for an arbitrarily anisotropic material. In the case of isotropy or cubic symmetry the constitutive equations take the form

$$D_i = \varepsilon E_i, \quad B_i = \mu H_i, \quad J_i = \sigma E_i. \tag{10.2.7}$$

Substituting (10.2.7) in Maxwell's vector equations (10.1.36) and (10.1.37), we obtain

$$\nabla \times \mathbf{H} = \frac{1}{c}[4\pi\sigma E + \varepsilon\dot{\mathbf{E}}], \tag{10.2.8}$$

$$\nabla \times \mathbf{E} = -\frac{1}{c}\mu\dot{\mathbf{H}}. \tag{10.2.9}$$

Now multiplying the time derivative of (10.2.8) by μ/c and adding the curl of (10.2.9), we obtain

$$\nabla \times \nabla \times \mathbf{E} + \frac{\mu}{c^2}[\varepsilon\ddot{\mathbf{E}} + 4\pi\sigma\dot{\mathbf{E}}] = 0, \tag{10.2.10}$$

which is a linear homogeneous vector equation in the vector field **E**, which yields wavelike solutions. A similar vector equation for the vector field **H** may readily be obtained. However, we will not bother to do so.

At this point it should be carefully noted that we did not need nor have we used either of the auxiliary scalar equations in (10.1.38), which with (10.2.7) yield

$$\nabla \cdot \mathbf{E} = \frac{4\pi\rho}{\varepsilon}, \quad \mu\nabla \cdot \mathbf{H} = 0, \tag{10.2.11}$$

to obtain our vector field equations on **E** (and **H**). That is why we use the word auxiliary. They are usually not needed to obtain solutions in the full electromagnetic case. Note that the first simply serves to define the charge density ρ after the vector field **E** has been determined, and the time derivative of the second is satisfied identically by virtue of (10.2.9).

10.3 Electromagnetism in Free-Space

Before discussing the electromagnetic field equations in free-space, we note that we have only a passing interest in the linear description presented in Sec. 10.2 because it is based on the linear constitutive assumptions in (10.2.1)–(10.2.3). On the other hand of particularly great interest to us is the case of free-space, for which the general equations (10.1.36)–(10.1.39) hold. In the case of free-space we have

$$\rho = 0, \quad \mathbf{J} = 0, \quad \mathbf{P} = 0, \quad \mathbf{M} = 0, \tag{10.3.1}$$

and Eqs. (10.1.36)–(10.1.38) with the aid of (10.1.39) take the respective forms

$$\nabla \times \mathbf{B} = \frac{1}{c}\frac{\partial \mathbf{E}}{\partial t}, \tag{10.3.2}$$

$$\nabla \times \mathbf{E} = -\frac{1}{c}\frac{\partial \mathbf{B}}{\partial t}, \tag{10.3.3}$$

$$\nabla \cdot \mathbf{E} = 0, \quad \nabla \cdot \mathbf{B} = 0, \tag{10.3.4}$$

which shows that in the general case Maxwell's equations in free-space become *linear*.

Now, first taking the time derivative of (10.3.2) and subtracting $\nabla \times$ of (10.3.3) and then taking the time derivative of (10.3.3) and adding $\nabla \times$ of (10.3.2), we obtain the respective forms

$$\nabla \times \nabla \times \mathbf{E} = -\frac{1}{c^2}\ddot{\mathbf{E}}, \tag{10.3.5}$$

$$\nabla \times \nabla \times \mathbf{B} = -\frac{1}{c^2}\ddot{\mathbf{B}}, \tag{10.3.6}$$

which with (10.3.4) and (7.3.8) yield, respectively, the two vector field equations

$$\nabla^2 \mathbf{E} = \frac{1}{c^2}\ddot{\mathbf{E}}, \quad \nabla^2 \mathbf{B} = \frac{1}{c^2}\ddot{\mathbf{B}}, \tag{10.3.7}$$

governing electromagnetic behavior in free-space. But these are just vector wave equations, with wave velocity c. Hence electromagnetic fields propagate in free-space (vacuum) with the velocity c, which heretofore has been the number of electric units in the magnetic unit of charge.

The scalar c then turns out to be identical with the speed of light in vacuum and, consequently, light is an electromagnetic disturbance with fields $\mathbf{E}$ and $\mathbf{B}$ governed by the wave equations (10.3.7).

We can say a bit more about the nature of the electromagnetic waves in free-space from the forms in (10.3.2)–(10.3.4) and (10.3.7). Since plane waves of the form

$$\mathbf{E} = \mathbf{E}^0 e^{i\xi(\mathbf{n}\cdot\mathbf{r} - ct)}, \quad \mathbf{B} = \mathbf{B}^0 e^{i\xi(\mathbf{n}\cdot\mathbf{r} - ct)}, \tag{10.3.8}$$

are solutions of (10.3.7), where $\mathbf{E}^0$ and $\mathbf{B}^0$ represent the amplitudes of the wave, ξ is the propagation wavenumber, $\mathbf{n}$ is the wave normal and $\mathbf{r}$ denotes the position coordinate. Substituting from (10.3.8) into (10.3.4), we obtain

$$\mathbf{n}\cdot\mathbf{E}^0 = 0, \quad \mathbf{n}\cdot\mathbf{B}^0 = 0, \tag{10.3.9}$$

which shows that the wave is transverse, i.e., that the amplitudes $\mathbf{E}^0$ and $\mathbf{B}^0$ lie in the plane normal to the propagation direction $\mathbf{n}$. The further substitution of the forms in (10.3.8) into (10.3.2) and (10.3.3) yields the respective forms

$$\mathbf{n}\times\mathbf{B}^0 = -\mathbf{E}^0, \quad \mathbf{n}\times\mathbf{E}^0 = \mathbf{B}^0, \tag{10.3.10}$$

which shows that the transverse amplitudes of the plane wave are mutually orthogonal. It should be carefully noted that the existence of these plane wave solutions of the electromagnetic field equations in free-space depends in a very crucial way on the assumption made by Maxwell in generalizing the magnetostatic field equations in (10.1.19) to the dynamic case in (10.1.35) as well as on the Maxwell version of Faraday's law in (10.1.10).

10.4 Electromagnetic Boundary Conditions

The differential equations of electromagnetism in the general case are given by (10.1.36)–(10.1.39). To these we have to adjoin the boundary (or jump) conditions of electromagnetism. As in the earlier cases of electrostatics and magnetostatics we require the integral statements, from which the differential equations in (10.1.36)–(10.1.38) may be obtained when suitable differentiability and continuity conditions are satisfied, in order to obtain the boundary conditions. The integral statements corresponding, respectively, to (10.1.36)–(10.1.38) are

$$\oint_C \mathbf{H}\cdot d\mathbf{r} = \frac{4\pi}{c}\int_s \mathbf{n}\cdot\mathbf{J}\,dS + \frac{1}{c}\frac{\partial}{\partial t}\int_s \mathbf{n}\cdot\mathbf{D}\,dS, \tag{10.4.1}$$

$$\oint_C \mathbf{E}\cdot d\mathbf{r} = -\frac{1}{c}\frac{\partial}{\partial t}\int_s \mathbf{n}\cdot\mathbf{B}\,dS, \tag{10.4.2}$$

$$\int_S \mathbf{n}\cdot\mathbf{D}\,dS = 4\pi\int_V \rho\,dV, \quad \int_S \mathbf{n}\cdot\mathbf{B}\,dS = 0, \tag{10.4.3}$$

where C is *any stationary* closed curve surrounding an open surface s and S is *any stationary* closed surface surrounding a volume V. We may apply these integral forms to *any* discontinuity surface, but at present we restrict ourselves to *stationary* discontinuity surfaces only. Since the spatial region is stationary and $\mathbf{J}, \dot{\mathbf{D}}$ and $\dot{\mathbf{B}}$ remain bounded and only ρ can become singular and give rise to a surface charge density σ as in electrostatics, it is clear from

(10.4.1)–(10.4.3) and our previous work with surfaces of discontinuity in Chapter 3 that the general jump conditions for the dynamic case are

$$\mathbf{n} \times [\mathbf{H}] = 0, \quad \mathbf{n} \times [\mathbf{E}] = 0, \tag{10.4.4}$$

$$\mathbf{n} \cdot [\mathbf{D}] = 4\pi\sigma, \quad \mathbf{n} \cdot [\mathbf{B}] = 0, \tag{10.4.5}$$

all of which we have seen before.

CHAPTER 11

Energy and Momentum in the Electromagnetic Field

11.1 Electromagnetic Energy

We now seek the expressions for the energy stored in and the energy flux of the electromagnetic field. However, since the description has been generalized beyond electro- and magneto-statics, we can no longer use the fundamental definitions of force to define energy in a relatively straightforward manner. In the case of electromagnetism the only procedure available is to operate formally on the equations and determine a scalar relation that is *quadratic* in the field variables and interpret the various terms that occur in the equation in light of our knowledge of the static electric and magnetic descriptions. Of course, we require that any interpretation that we give in the electromagnetic case reduce to that which we know when we reduce to the electro- and magneto-static cases.

In accordance with the foregoing we consider the electromagnetic equations in (10.1.36) and (10.1.37), which in indicial notation take the respective forms

$$e_{ijk}H_{k,j}=\frac{4\pi}{c}J_i+\frac{1}{c}\dot{D}_i, \tag{11.1.1}$$

$$e_{ijk}E_{k,j}=-\frac{1}{c}\dot{B}_i, \tag{11.1.2}$$

and contract (11.1.1) with E_i and (11.1.2) with H_i and subtract the second from the first to obtain

$$e_{ijk}H_{k,j}E_i-e_{ijk}E_{k,j}H_i=\frac{4\pi}{c}E_iJ_i+\frac{1}{c}E_i\dot{D}_i+\frac{1}{c}H_i\dot{B}_i,$$

which readily enables us to write

$$-e_{jik}(E_iH_k)_{,j}=\frac{1}{c}(E_i\dot{D}_i+H_i\dot{B}_i)+\frac{4\pi}{c}E_iJ_i. \tag{11.1.4}$$

Multiplying through by $c/4\pi$ and converting to vector notation, we obtain

$$-\frac{c}{4\pi}\nabla\cdot(\mathbf{E}\times\mathbf{H})=\frac{1}{4\pi}(\mathbf{E}\cdot\dot{\mathbf{D}}+\mathbf{H}\cdot\dot{\mathbf{B}})+\mathbf{E}\cdot\mathbf{J}. \tag{11.1.5}$$

Integrating over an arbitrary volume and applying the divergence theorem, we find

$$-\frac{c}{4\pi}\int_S \mathbf{n}\cdot\mathbf{E}\times\mathbf{H}\,dS=\frac{1}{4\pi}\int_V(\mathbf{E}\cdot\dot{\mathbf{D}}+\mathbf{H}\cdot\dot{\mathbf{B}})\,dV+\int_V \mathbf{E}\cdot\mathbf{J}\,dV. \tag{11.1.6}$$

Equation (11.1.6) is Poynting's theorem. It is nothing more than a mathematical consequence of Maxwell's equations, so far. We will now try to interpret the terms physically. When we do this we will have an electromagnetic energy identity, which could be called the electromagnetic energy equation. It is not the conservation of energy equation when there is *any* interaction present such as, e.g., thermal or mechanical. However, in any event, it is always a mathematical identity which is compatible with Maxwell's equations (10.1.36) and (10.1.37).

Since we are interested in the nonlinear case we will transform Poynting's theorem to a form in which the terms can be interpreted in the general nonlinear case. To this end we substitute (10.1.39) into (11.1.6) to obtain

$$-\frac{c}{4\pi}\int_S \mathbf{n}\cdot\mathbf{E}\times\mathbf{H}\,dS=\frac{1}{4\pi}\int_V[\mathbf{E}\cdot(\dot{\mathbf{E}}+4\pi\dot{\mathbf{P}})+(\mathbf{B}-4\pi\mathbf{M})\cdot\dot{\mathbf{B}}]\,dV+\int_V \mathbf{E}\cdot\mathbf{J}\,dV, \tag{11.1.7}$$

which enables us to write

$$-\frac{c}{4\pi}\int_S \mathbf{n}\cdot\mathbf{E}\times\mathbf{H}\,dS=\frac{\partial}{\partial t}\int_V\frac{1}{8\pi}(\mathbf{E}\cdot\mathbf{E}+\mathbf{B}\cdot\mathbf{B})\,dV$$

$$+\int_V[\mathbf{E}\cdot\dot{\mathbf{P}}-\mathbf{M}\cdot\dot{\mathbf{B}}+\mathbf{E}\cdot\mathbf{J}]\,dV. \tag{11.1.8}$$

We are now in a position to interpret the individual terms in (11.1.8). To this end let us define $\mathbf{h}$ by

$$\mathbf{h}=\frac{c}{4\pi}\mathbf{E}\times\mathbf{H}, \tag{11.1.9}$$

and note that the first volume integral on the right-hand side of (11.1.8) is the electric field energy density from electrostatics plus the magnetic field energy density from magnetostatics. Let us denote this density by U^F and call it the electromagnetic field energy density. Then, we have

$$U^F=\frac{1}{8\pi}(\mathbf{E}\cdot\mathbf{E}+\mathbf{B}\cdot\mathbf{B}), \tag{11.1.10}$$

in Gaussian units, and with (11.1.9) and (11.1.10), (11.1.8) may be written in the form

$$-\int_S \mathbf{n}\cdot\mathbf{h}\,dS = \frac{\partial}{\partial t}\int_V U^F\,dV + \int_V [\mathbf{E}\cdot\dot{\mathbf{P}} - \mathbf{M}\cdot\dot{\mathbf{B}} + \mathbf{E}\cdot\mathbf{J}]\,dV, \quad (11.1.11)$$

in general. Equation (11.1.11) may now be interpreted in the following way: The time rate of change of electromagnetic field energy in an *arbitrary* volume is equal to *minus* the rate at which electromagnetic energy flows out of the surface enclosing that volume minus the rate at which the electric field and magnetic induction field supply energy to (do work on) the matter within the *arbitrary* volume. This interpretation holds in general. Clearly, it means that the vector **h** represents the energy flux vector across a surface, which is positive when outwardly directed. It is called the Poynting vector, and it has the units of energy per unit area per unit time or power per unit area. Equation (11.1.11) is the electromagnetic energy identity under all circumstances, but it is *not* the equation of the conservation of energy because there are other types of energy present, which are interacting by virtue of the interaction terms in brackets.

In the case of linear electromagnetism discussed in Sec. 10.2 we have

$$P_i = \chi_{ij}E_j, \quad M_i = \zeta_{ij}B_j, \quad (11.1.12)$$

which enables (11.1.11) to be written in the form

$$-\int_S n_i h_i\,dS = \int_V \left[\frac{1}{4\pi}(E_i\dot{E}_i + \mathrm{B}_i\dot{\mathrm{B}}_i) + E_i\chi_{ij}\dot{E}_j - \dot{B}_i\zeta_{ij}B_j + E_iJ_i\right]dV, \quad (11.1.13)$$

from which we obtain

$$-\int_S n_i h_i\,dS = \frac{1}{2}\frac{\partial}{\partial t}\int_V \left[E_i\left(\frac{\delta_{ij}}{4\pi} + \chi_{ij}\right)E_j + B_j\left(\frac{\delta_{ij}}{4\pi} - \zeta_{ij}\right)B_i\right]dV + \int_V E_iJ_i\,dV. \quad (11.1.14)$$

Substituting from (11.1.12) and (10.1.39) into (11.1.14), we obtain

$$-\int_S n_i h_i\,dS = \frac{\partial}{\partial t}\int_V \frac{1}{8\pi}[E_iD_i + B_iH_i]\,dV + \int_V E_iJ_i. \quad (11.1.15)$$

Since the first volume integral on the right-hand side of (11.1.15) is the electromagnetic energy including the energy of interaction of the electric and magnetic fields with the matter in the linear case, we may write (11.1.15) in the form

$$-\int_S n_i h_i\,dS = \frac{\partial}{\partial t}\int_V U^{EM}\,dV + \int_V E_iJ_i, \quad (11.1.16)$$

where

$$U^{EM} = \frac{1}{8\pi}[E_iD_i + H_iB_i], \quad (11.1.17)$$

is defined as the electromagnetic energy density in this linear case. We identify this term as the electromagnetic energy density in the linear case because it is differentiated with respect to time in (11.1.16). The energy density U^{EM} in (11.1.17) consists of what we have heretofore called the field energy U^F in (11.1.10) plus the terms that arise from the rates of interaction of the fields with the matter in (11.1.11). This combination is possible only in the linear case.

Thus in this special linear case we can give the classical interpretation of Poynting's theorem if the only other type of energy is thermal and it is excluded from the system. Under this circumstance the E_iJ_i term represents the usual Joule heat (dissipation). Poynting's equation (11.1.16) in this purely linear electromagnetic case now asserts that the time rate of increase of internal (electromagnetic) energy in an arbitrary volume is equal to minus the rate at which electromagnetic energy flows out of the surface enclosing the volume minus the rate at which electric energy is dissipated inside the volume by thermal means. Before proceeding further we note, as we have previously, that in this treatment we have no particular interest in linear electromagnetic theory and, consequently, no particular interest in the *interpretation* of Poynting's theorem in the linear electromagnetic case.

11.2 Electromagnetic Momentum and Force

In this section we consider the mechanical body forces of electromagnetic origin. However, for historical reasons and because of a certain lack of clarity we note that this is not terribly good terminology. We now rephrase the statement in the following way: We are interested in the rate at which the electromagnetic field supplies mechanical linear momentum to the matter at the point in question.[6] But this is just the electromagnetic body force density **f**. The main reason for the restatement is that the local portion of the force of macroscopic electromagnetic origin will be included in the mechanical traction vector because these local portions were excluded from the definition of the field vectors **E** and **B** appearing in Maxwell's equations. Another reason for the restatement is that the sources of the fields producing the forces may equally readily be within the body as without.

In developing electrostatics and magnetostatics we took as fundamental, i.e., as the starting point, the definitions of electric and magnetic forces and appropriately generalized the definitions where required in developing the field theory. We now take the natural course of *assuming* that the expression for the electromagnetic body force density, i.e., the rate of supply of mechanical linear momentum from the electromagnetic field to the matter, is simply the sum of the electric plus the magnetic body force densities, $\mathbf{f}^E$ and $\mathbf{f}^M$, which from Eqs. (4.1.11) and (8.1.15), respectively, are given by

$$\mathbf{f}^E = \rho\mathbf{E} + \mathbf{P}\cdot\nabla\mathbf{E}, \tag{11.2.1}$$

$$\mathbf{f}^M = \mathbf{J}^M \times \mathbf{B} + \mathbf{M}\cdot(\mathbf{B}\nabla) = \frac{1}{c}\mathbf{J}^E \times \mathbf{B} + \mathbf{M}\cdot(\mathbf{B}\nabla), \tag{11.2.2}$$

where the c has been introduced because the current is now in electrostatic units. However, we must remember that although the charge equation of electrostatics remains intact in electromagnetism, the Ampere equation of magnetostatics has been generalized from Eq. (7.3.12), which is of the form

$$\nabla \times \mathbf{H} = \frac{4\pi}{c}\mathbf{J}, \tag{11.2.3}$$

to Eq. (10.1.35), which is of the form

$$\nabla \times \mathbf{H} = \frac{4\pi}{c}\mathbf{J} + \frac{1}{c}\dot{\mathbf{D}} = \frac{4\pi}{c}(\mathbf{J} + \dot{\mathbf{P}}) + \frac{1}{c}\dot{\mathbf{E}}, \tag{11.2.4}$$

by Maxwell. Clearly, $\mathbf{J}$ and $\dot{\mathbf{P}}$ belong to the matter whereas $\dot{\mathbf{E}}$ belongs to space. It is clear from the form of the generalization that $\dot{\mathbf{P}}$ behaves as a material current in exactly the same way as $\mathbf{J}$. Hence, it seems clear from the development of magnetostatics and this nonsteady version of Ampere's law that $\dot{\mathbf{P}}$ should be included in the same manner as $\mathbf{J}$ in the force equation as well as in the field equations. This is not true for $\dot{\mathbf{E}}$ because it belongs to the space and not the matter. Thus to the magnetic force in the nonsteady state we must add the term $\dot{\mathbf{P}} \times \mathbf{B}/c$. Accordingly, with this addition and (11.2.1) and (11.2.2) the expression for the total force $\mathbf{f} = \mathbf{f}^E + \mathbf{f}^M$ may be written in the form

$$\mathbf{f} = \rho\mathbf{E} + \mathbf{P}\cdot\nabla\mathbf{E} + \frac{1}{c}\mathbf{J} \times \mathbf{B} + \mathbf{M}\cdot(\mathbf{B}\nabla) + \frac{1}{c}\dot{\mathbf{P}} \times \mathbf{B}, \tag{11.2.5a}$$

which in indicial notation becomes

$$f_i = \rho E_i + P_k E_{i,k} + \frac{1}{c} e_{ijk} J_j B_k + M_k B_{k,i} + \frac{1}{c} e_{ijk} \dot{P}_j B_k. \tag{11.2.5b}$$

Now that we have the expression (11.2.5) for the body force, we are ready to employ Maxwell's equations (10.1.36)–(10.1.38), which for convenience we rewrite here in indicial notation,

$$e_{lmn} H_{n,m} = \frac{4\pi}{c} J_l + \frac{1}{c} \dot{D}_l, \tag{11.2.6}$$

$$e_{lmn} E_{n,m} = -\frac{1}{c} \dot{B}_l, \tag{11.2.7}$$

$$D_{m,m} = 4\pi\rho, \quad B_{m,m} = 0, \tag{11.2.8}$$

and the definitions in (10.1.39), which we write in the form

$$D_r = E_r + 4\pi P_r, \quad B_r = H_r + 4\pi M_r. \tag{11.2.9}$$

Substituting from (11.2.6), $(11.2.8)_1$ and $(11.2.9)_1$ into (11.2.5b), we obtain

$$f_i = \frac{D_{k,k}}{4\pi} E_i + \left(\frac{D_k - E_k}{4\pi}\right) E_{i,k} + \frac{1}{c} e_{ijk} \left[\frac{c}{4\pi} e_{jmn} H_{n,m} - \frac{\dot{D}_j}{4\pi}\right] B_k$$

$$+ M_k B_{k,i} + \frac{1}{c} e_{ijk} \dot{P}_j B_k, \tag{11.2.10}$$

which with the tensor identity used earlier,[22] (11.2.9) and $(11.2.8)_2$ may be written in the form

$$f_i = \frac{1}{4\pi}[D_k E_i + H_j B_k]_{,k} - \frac{1}{8\pi}(H_k H_k)_{,i} + 2\pi(M_k M_k)_{,i}$$

$$- \frac{1}{4\pi}\left[E_k E_{i,k} + \frac{1}{c} e_{ikl} \dot{E}_k B_l\right]. \tag{11.2.11}$$

In order to carry this further we digress and operate on both sides of (11.2.7) with e_{lrs} to obtain

$$e_{lrs} e_{lmn} E_{n,m} = -\frac{1}{c} e_{lrs} \dot{B}_l, \tag{11.2.12}$$

which with the same tensor idensity[22] just employed yields

$$E_{s,r} - E_{r,s} = -\frac{1}{c} e_{lrs} \dot{B}_l. \tag{11.2.13}$$

Changing dummy indices, from (11.2.13) we may write

$$E_k E_{i,k} = \tfrac{1}{2}(E_k E_k)_{,i} - E_k \frac{1}{c} e_{lki} \dot{B}_l. \tag{11.2.14}$$

The substitution of (11.2.14) into (11.2.11) with a further change of dummy indices yields

$$f_i = \frac{1}{4\pi}[D_k E_i + B_k H_i - \tfrac{1}{2}(E_j E_j + H_j H_j - 16\pi^2 M_j M_j)\delta_{ik}]_{,k} - \frac{1}{4\pi c} e_{ikl}[E_k \dot{B}_l + \dot{E}_k B_l],$$

which may be written in the final form

$$f_i = \frac{1}{4\pi}[D_k E_i + B_k H_i - \tfrac{1}{2}(E_j E_j + H_j H_j - 16\pi^2 M_j M_j)\delta_{ik}]_{,k} - \frac{\partial}{\partial t}\left[\frac{1}{4\pi c} e_{ikl} E_k B_l\right]. \tag{11.2.15}$$

The expression within the first set of brackets in (11.2.15) is a second rank tensor and is called the Maxwell electromagnetic stress tensor T_{ki}^{EM}. Hence, we may write

$$T_{ki}^{EM} = \frac{1}{4\pi}[D_k E_i + B_k H_i - \tfrac{1}{2}(E_j E_j + H_j H_j - 16\pi^2 M_j M_j)\delta_{ik}], \tag{11.2.16}$$

for the Maxwell electromagnetic stress tensor, which enables (11.2.15) to be written in the form

$$f_i = T^{EM}_{ki,k} - \frac{\partial}{\partial t}\left[\frac{1}{4\pi c} e_{ikl} E_k B_l\right], \tag{11.2.17}$$

or in invariant dyadic form (11.2.16) may be written

$$\mathbf{T}^{EM} = \frac{1}{4\pi}[\mathbf{DE} + \mathbf{BH} - \tfrac{1}{2}(\mathbf{E}\cdot\mathbf{E} + \mathbf{H}\cdot\mathbf{H} - 16\pi^2\mathbf{M}\cdot\mathbf{M})\mathbf{I}], \tag{11.2.18}$$

and (11.2.17) may be written

$$\mathbf{f}^{EM} = \nabla\cdot\mathbf{T}^{EM} - \frac{\partial}{\partial t}\left(\frac{1}{4\pi c}\mathbf{E}\times\mathbf{B}\right). \tag{11.2.19}$$

The quantity differentiated with respect to time in (11.2.19) is called the electromagnetic momentum, and we give it the symbol **g**. Thus from (11.2.19) we have

$$\mathbf{f}^{EM} = \nabla\cdot\mathbf{T}^{EM} - \frac{\partial}{\partial t}\mathbf{g}, \tag{11.2.20}$$

where

$$\mathbf{g} = \frac{1}{4\pi c}\mathbf{E}\times\mathbf{B}. \tag{11.2.21}$$

These last relations in (11.2.19) and (11.2.20) are very important and very interesting. For one thing the electromagnetic field possesses linear momentum *even in free-space*. Moreover, from (11.1.9), (11.2.9) and (11.2.21) in *free-space* we have the relation

$$\mathbf{g} = \frac{1}{c^2}\mathbf{h} \tag{11.2.22}$$

between the electromagnetic momentum **g** and the Poynting energy flux vector **h**. However, in matter according to this description the Poynting vector **h** defined in (11.1.9) and the electromagnetic momentum vector **g** defined in (11.2.21) are not necessarily co-linear. The integral form of the electromagnetic momentum equation may be obtained by integrating (11.2.20) over an arbitrary volume V and employing the divergence theorem and is given by

$$\int_V \mathbf{f}^{EM}\,dV = \int_S \mathbf{n}\cdot\mathbf{T}^{EM}\,dS - \frac{\partial}{\partial t}\int_V \mathbf{g}\,dV. \tag{11.2.23}$$

Equation (11.2.23) says that the time rate of change of electromagnetic momentum in an *arbitrary* volume is equal to the force exerted on the surface of the volume by the electromagnetic traction vector minus the rate of supply of mechanical linear momentum from the electromagnetic field to the matter. The similarity between the momentum equation in (11.2.23) and the electromagnetic energy equation in (11.1.11) is quite clear. Among the equations presented in Secs. 11.1 and 11.2 are the general forms which are useful in the description of the nonlinear interaction of the electromagnetic field with a deformable continuum.

As usual, the integral forms are more fundamental than the differential forms which may be derived from them when appropriate differentiability conditions are assumed. The fact that electromagnetic momentum exists when there are **E** and **B** vectors in free space is important for the existence of a light photon which has momentum and no mass. It should be noted that the electromagnetic momentum balance relations in (11.2.20) could have been obtained more easily if we had simply assumed that the electromagnetic Maxwell tensor in (11.2.18) was the sum of the electrostatic and magnetostatic stress tensors in (4.2.9) and (8.2.7) instead of assuming the expressions for the forces. Then we would have found that $\dot{\mathbf{P}}$ had a body force density associated with it in the same way as **J**, instead of having to realize that Maxwell's extension of the Ampere equation implied such a term at the outset. However, since our starting point has consistently been the expressions for force and force density, it seemed more logical to proceed in the same way here especially since it seems to provide more physical insight. At this point we mention in passing that all field vectors and all equations are with respect to a stationary (or inertial) reference system. We will return to this point in some detail a little later on after we consider a few other questions which still remain before us.

CHAPTER 12

The Influence of Motion on the Electromagnetic Field Equations

12.1 Boundary Conditions at Moving Surfaces of Discontinuity

We have already obtained the electromagnetic jump conditions across stationary surfaces of discontinuity in Sec. 10.4. But since we are ultimately interested in considering moving deformable media, we will need the electromagnetic jump conditions at *moving* surfaces of discontinuity. As we shall see the fact that the surfaces are moving alters the jump conditions. Consider the surface $\mathscr{S}$ moving with velocity $\mathbf{v}$ shown in Figure 32. The surface has unit normal $\mathbf{n}$ and two orthogonal unit vectors $\boldsymbol{\tau}$ and $\mathbf{N}$ normal to $\mathbf{n}$, but otherwise arbitrary and satisfying

$$\boldsymbol{\tau} = \mathbf{n} \times \mathbf{N}. \tag{12.1.1}$$

In Figure 32 the arbitrary counterclockwise circuit C is normal to $\mathbf{N}$. The jump conditions on $\mathbf{H}$ and $\mathbf{E}$ may be obtained from the fundamental integral forms in (10.4.1) and (10.4.2), which for the circuit C take the respective forms

$$\oint_C \mathbf{H}\cdot d\mathbf{r} = \frac{4\pi}{c}\int_a \mathbf{N}\cdot\mathbf{J}\,dS + \frac{1}{c}\frac{\partial}{\partial t}\int_a \mathbf{N}\cdot\mathbf{D}\,dS, \tag{12.1.2}$$

$$\oint_C \mathbf{E}\cdot d\mathbf{r} = -\frac{1}{c}\frac{\partial}{\partial t}\int_a \mathbf{N}\cdot\mathbf{B}\,dS. \tag{12.1.3}$$

We now apply (12.1.2) and (12.1.3) to the arbitrary path denoted C^+ and C^- surrounding $\mathscr{S}$, as shown in Figure 33, and take the limit as C^+ and C^- both approach s on $\mathscr{S}$ while the areas a^+ and a^- vanish so that the length ℓ of s remains finite in the limit. The left-hand time-independent integrals in (12.1.2) and (12.1.3) are treated in exactly the same manner as Eq. (3.5.13) was in Chapter 3 on electrostatics, but now since we have a nonzero right-hand side we briefly repeat the procedure here.

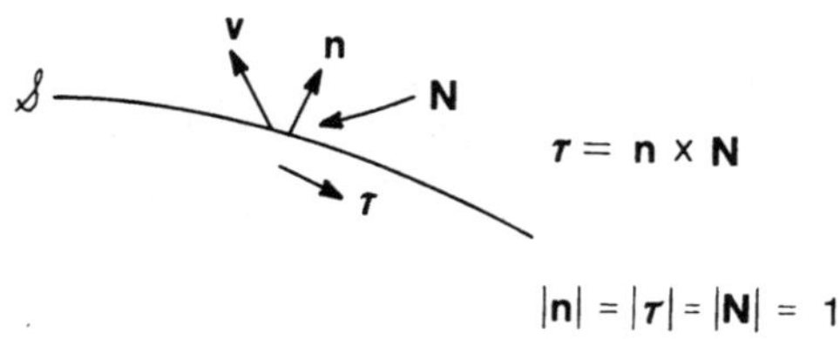

FIGURE 32. Moving surface of discontinuity.

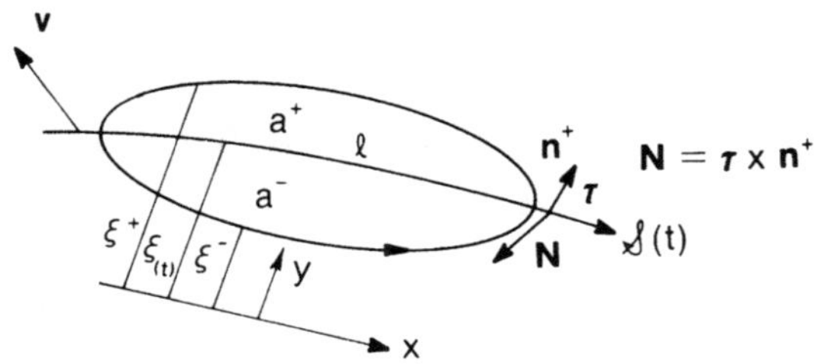

FIGURE 33. Coordinates and regions for the determination of jump conditions across the moving surface of discontinuity.

The integral on the left-hand side of Eq. (12.1.2) takes the form

$$\oint_C \mathbf{H}\cdot d\mathbf{r} = \lim_{\substack{a^+\to 0\\ a^-\to 0}} \int_{C^-} \mathbf{H}^-\cdot d\mathbf{r} + \int_{C^+} \mathbf{H}^+\cdot d\mathbf{r} = \int_l \boldsymbol{\tau}\cdot(\mathbf{H}^- - \mathbf{H}^+)\,ds, \quad (12.1.4)$$

which with (12.1.1) in the limit yields

$$\oint_C \mathbf{H}\cdot d\mathbf{r} = \int_l \mathbf{N}\cdot\mathbf{n}\times(\mathbf{H}^+ - \mathbf{H}^-)\,ds. \quad (12.1.5)$$

Since **J** remains bounded the first integral on the right-hand side of (12.1.2) vanishes in the limit $a \to 0$. But the second integral on the right-hand side of (12.1.2) has the time derivative outside the integral sign and must be discussed carefully because the motion of $\mathscr{S}$ gives rise to a term on account of the discontinuity in **D**. To this end we write

$$\lim_{\substack{a^+\to 0\\ a^-\to 0}} \frac{\partial}{\partial t}\int_a \mathbf{N}\cdot\mathbf{D}\,dS = \frac{\partial}{\partial t}\int_{a^+} \mathbf{N}\cdot\mathbf{D}\,dS + \frac{\partial}{\partial t}\int_{a^-} \mathbf{N}\cdot\mathbf{D}\,dS, \quad (12.1.6)$$

which with the aid of the x, y coordinate system shown in Figure 33 may be written in the form

$$\lim_{\substack{a^+\to 0\\ a^-\to 0}} \frac{\partial}{\partial t}\int_a \mathbf{N}\cdot\mathbf{D}\,dS = \frac{\partial}{\partial t}\int_l dx \int_{\xi(t)}^{\xi^+} D_N\,dy + \frac{\partial}{\partial t}\int_l dx \int_{\xi^-}^{\xi(t)} D_N\,dy. \quad (12.1.7)$$

Employing Leibniz' rule for differentiation of the integrals, we obtain

$$\lim_{\substack{a^+\to 0\\ a^-\to 0}} \frac{\partial}{\partial t}\int_a \mathbf{N}\cdot\mathbf{D}\,dS = \int_l dx\left[\left[\int_{\xi(t)}^{\xi^+} \frac{\partial D_N}{\partial t}dy + \int_{\xi^-}^{\xi(t)} \frac{\partial D_N}{\partial t}dy\right]\right.$$

$$\left. + \int_l dx[-\dot{\xi}(t)D_N^+(\xi(t)) + \dot{\xi}(t)D_N^-(\xi(t))]\right]. \qquad (12.1.8)$$

Since $\partial D_N/\partial t$ remains bounded and $\dot{\xi}(t) = v_n = \mathbf{n}\cdot\mathbf{v}$, in the limit (12.1.8) yields

$$\lim_{\substack{a^+\to 0\\ a^-\to 0}} \frac{\partial}{\partial t}\int_a \mathbf{N}\cdot\mathbf{D}\,dS = 0 + \int_l v_n(D_N^- - D_N^+)\,dx = \int_l \mathbf{N}\cdot(\mathbf{D}^- - \mathbf{D}^+)v_n\,dx. \qquad (12.1.9)$$

Substituting from (12.1.5) and (12.1.9) into (12.1.2) and utilizing the fact that the integral containing $\mathbf{J}$ vanishes in the limit, we obtain

$$\int_l \mathbf{N}\cdot\left[\mathbf{n}\times(\mathbf{H}^+ - \mathbf{H}^-) + \frac{v_n}{c}(\mathbf{D}^+ - \mathbf{D}^-)\right]ds = 0, \qquad (12.1.10)$$

which holds for any arbitrary l and accompanying $\mathbf{N}$ on $\mathscr{S}$. Hence we have

$$\mathbf{n}\times[\mathbf{H}] + \frac{v_n}{c}[\mathbf{D}_t] = 0, \qquad (12.1.11)$$

as the jump condition across a surface of discontinuity $\mathscr{S}$ moving with normal velocity $v_n = \mathbf{n}\cdot\mathbf{v}$ and where $\mathbf{D}_t$ denotes the tangential component of $\mathbf{D}$. Clearly, in the same manner, from Faraday's law (12.1.3) we obtain the jump condition

$$\mathbf{n}\times[\mathbf{E}] - \frac{v_n}{c}[\mathbf{B}_t] = 0. \qquad (12.1.12)$$

Clearly, at a moving surface of discontinuity $\mathscr{S}$ with velocity $\mathbf{v}$, the integral forms in (10.4.3), which are

$$\int_S \mathbf{n}\cdot\mathbf{D}\,dS = 4\pi\int_V \rho\,dV, \qquad (12.1.13)$$

$$\int_S \mathbf{n}\cdot\mathbf{B}\,dS = 0, \qquad (12.1.14)$$

yield the same jump conditions as in the static case because no time derivatives appear in these integral relations. Thus in the dynamic case we still have the jump conditions in (3.5.11) and (7.5.5) which are

$$\mathbf{n}\cdot[\mathbf{D}] = 4\pi\sigma, \qquad \mathbf{n}\cdot[\mathbf{B}] = 0. \qquad (12.1.15)$$

12.2 Relative Motion and the Speed of Light

We now have Maxwell's equations referred to a *stationary* (inertial) coordinate system and the jump conditions across a surface of discontinuity moving with respect to our stationary coordinate system, shown in Figure 34. The scalar ρ and the vector fields **E**, **B**, **P**, **M**, **J** appearing in Maxwell's equations are referred to the stationary system S and the spatial differentiations in Maxwell's equations are with respect to the coordinates x_k of the stationary system. Now, from our point of view the fields **E** and **B** belong to the space and although we are interested in their transformation properties, our primary interest here is with respect to our stationary system S, which is the way we already have them. But ρ, **P**, **M** and **J** belong to the *matter* and their values with respect to a coordinate system moving with the velocity of the matter at the point P, at which they exist in the matter, might very well be different than their values with respect to our stationary system. In order to determine their expression in our stationary system we must investigate the question of the transformation of the field vectors between coordinate systems in relative motion.

We begin the aforementioned investigation by considering the Galilean transformations of classical mechanics, which transform us from one inertial system to another. To this end we note that the equations of classical mechanics in any one inertial coordinate system take the form

$$\sum \mathbf{f} = m\ddot{\mathbf{r}}. \tag{12.2.1}$$

If we then have another coordinate system which is translating with uniform velocity $\mathbf{v}_0$ relative to the first, and the position coordinate in the second system is R, we have

$$\mathbf{R} = \mathbf{r} - (\mathbf{v}_0 t + \mathbf{b}), \tag{12.2.2}$$

where **b** denotes the relative displacement of the two coordinate systems at $t = 0$. Clearly, if we let **F** represent force and M mass in the second system, we have

$$\ddot{\mathbf{R}} = \ddot{\mathbf{r}}, \quad \mathbf{F} = \mathbf{f}, \quad M = m \tag{12.2.3}$$

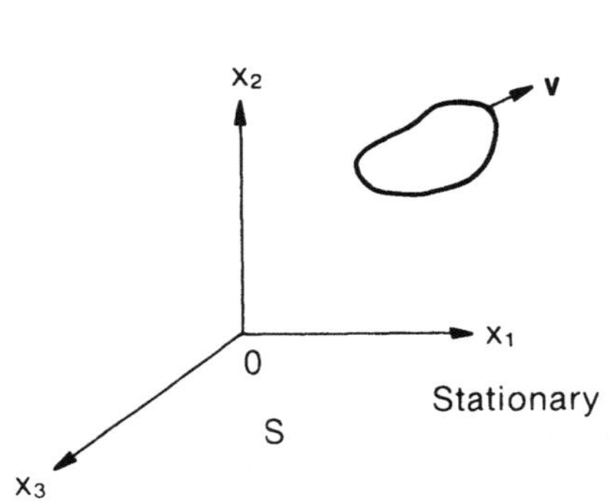

FIGURE 34. Moving body and stationary coordinate system.

since from (12.2.1) and (12.2.2), we have

$$\sum \mathbf{F} = M\ddot{\mathbf{R}} = \sum \mathbf{f} = m\ddot{\mathbf{R}} = \sum \mathbf{f} = m\ddot{\mathbf{r}}, \tag{12.2.4}$$

which shows that the mechanical equations of motion are *invariant* under such a change of frame. These *particular* frames of reference in which Newton's law of motion holds are called inertial frames. Newton's law of motion does not hold in other (noninertial) frames. In other words the mechanical equations of motion are not *invariant* under more general time-dependent (noninertial, non-Galilean) transformations. This notion of *invariance* is very important and very useful physically.

Now, it turns out that Maxwell's equations (10.1.36)–(10.1.38) are not invariant under Galilean transformations. In other words if Maxwell's equations hold in one *inertial* coordinate system, the equations obtained from them, when written in terms of the coordinates of *another inertial* system obtained from the first by a Galilean transformation, will not have the *same form* as the equations in the first *inertial* coordinate system, i.e., as Maxwell's equations. This is one very important fact concerning the electromagnetic equations. Another important fact concerning these equations, which we already know, is that they govern the propagation of light in vacuum with speed c relative to the *inertial* (fixed) coordinate system in which they are written. Since c is not infinite it follows that light propagates as a wave disturbance with finite speed.

As we know, if we move with respect to, say, a string in which a wave is propagating with speed s, we would measure (observe) a relative velocity, say, $s - u$, where u is our speed relative to the string in the direction in which the wave is propagating. The same sort of result would be obtained with any other wave motion with which we are familiar such as a water wave. Consequently, it seems logical to expect this same sort of situation to persist in the case of the light wave, since it also is a propagating wave the same as the others. This quite logical sort of reasoning led the physicists of the last century and the beginning of this one to look for *the* rest (or inertial) system for the light (Maxwell's equations). This system was called the ether. They looked for this system by trying to measure differences in the speed of light relative to fixed (with respect to the earth) and moving sources and detectors. All results were negative. No difference in the measured speed was detected. But the speed of light is very very fast and the lack of any observed differences could be attributed to lack of sufficient accuracy in the measuring apparatus.

Finally, one such experiment, the Michelson and Morley experiment,[28] was considered to be sufficiently accurate to have detected such a very small difference if—indeed—one existed. The result of this experiment was negative also. This is the third important fact concerning light (Maxwell's equations). Everything was up in the air concerning this state of events when Einstein independent of the Michelson and Morley experiment, with which he was not familiar, asserted on purely theoretical grounds that the speed of light

c in vacuum was *independent* of the *relative* motion of the source *with respect* to the detector. This by itself is a tremendous departure from all human experience with wave motion, as we already discussed.

Since Maxwell's equations describe the propagation of light in vacuum with respect to a particular *inertial* system, namely our *fixed* laboratory system, they must also describe the propagation of light in vacuum with respect to any other *inertial* reference system, e.g., any of the ones used in the Michelson and Morley experiment, i.e., one moving with constant uniform velocity with respect to our *fixed* system. This means that Maxwell's equations must have the same form in *any* inertial system.

Einstein took it upon himself to determine the transformation from one inertial system to another which would leave Maxwell's equations invariant. He did this independently of Maxwell's equations by considering a spherical light source emitting into vacuum in two different inertial coordinate systems, which coincide as the light is emitted, and imposing the condition that the speed of the spherical wave front be c in *both* inertial systems.[7] In doing this we consider relative motion of the frames in the x_1-direction only, as shown in Figure 35, where $\mathbf{v}$ is the relative velocity of the primed frame with respect to the unprimed frame. Since, by physical hypothesis, only the relative translational motion of the two inertial frames counts, the relative velocity of the unprimed frame with respect to the primed frame must be $-\mathbf{v}$. Since, by hypothesis, *all* physical laws are to be of the same form in both the primed and unprimed coordinate systems in such a way that no special physical preference be given to the origin of a coordinate system or any other point in that system, the description of an *event* in one inertial coordinate system must be a linear function of the specification of the *same event* in another inertial coordinate system. In other words we must require the coordinates (x'_k, t') to be a linear function of the coordinates (x_k, t).

We will now deduce the transformation law for the very special case we are considering of a uniform relative translational motion in the x_1 direction

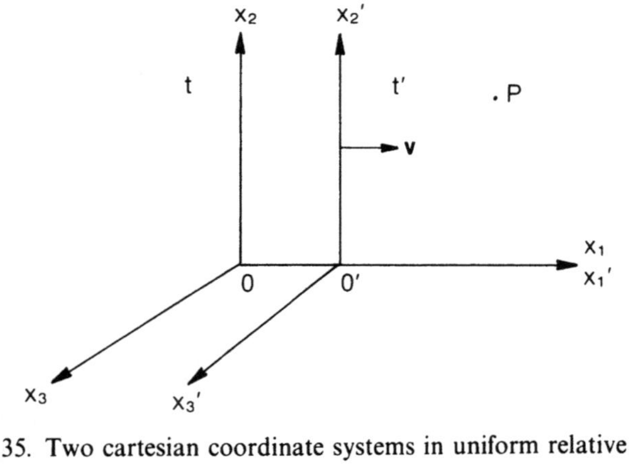

FIGURE 35. Two cartesian coordinate systems in uniform relative motion.

of the frames with relative speed v. We further require the origin of the unprimed system at $t = 0$ to coincide with the origin of the primed system at $t' = 0$. This denotes the instant of emission of the light. Since by assumption the x_1 and x_1' axes coincide throughout the process—this seeming physically plausible—we must have the planes $x_2 = 0$, $x_3 = 0$ corresponding, respectively, with the planes $x_2' = 0$, $x_3' = 0$. Since, as we have already stated, no particular point of any coordinate system is to be preferred over any other point, all planes parallel to these origin planes in the unprimed system must be parallel in the primed system. Moreover, it is clear physically that no preference should be given to the 2-direction over the 3-direction. Clearly then from these considerations we can have at most

$$x_2' = \varkappa(v)x_2, \quad x_3' = \varkappa(v)x_3, \tag{12.2.5}$$

where $\varkappa$ can be a function of v at most. If now we reverse the directions of the x_1 and x_3 axes as well as of the x_1' and x_3' axes we would have the situation shown in Figure 36, from which it is clear that the coordinates x_2 and x_2' still determine the same plane. However, now the two inertial systems have exchanged roles; the unprimed system moving with velocity v relative to the primed system in the $+x_1(x_1')$ direction. Hence, we may conclude as before that

$$x_2 = \varkappa(v)x_2'. \tag{12.2.6}$$

The substitution of (12.2.6) in (12.2.5) reveals that

$$\varkappa(v)\varkappa(v) = 1, \tag{12.2.7}$$

and since the positive directions of the x_2 and x_2' axes are the same, we must have

$$\varkappa(v) = 1. \tag{12.2.8}$$

Clearly, similar remarks hold for the 3-direction and we have

$$x_2' = x_2, \quad x_3' = x_3. \tag{12.2.9}$$

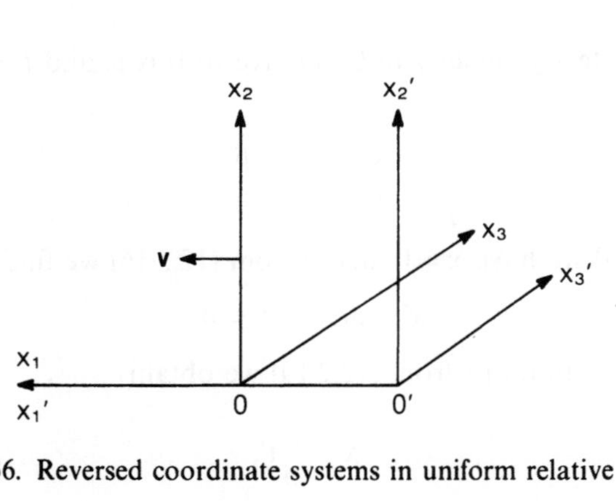

FIGURE 36. Reversed coordinate systems in uniform relative motion.

Now, the equation of the surface of the spherical light wave, which was emitted as stated, as measured in the unprimed system is

$$x_k x_k = c^2 t^2, \tag{12.2.10}$$

and in the primed system is

$$x'_l x'_l = c'^2 t'^2, \tag{12.2.11}$$

but according to the main assertion of the theory of relativity $c' = c$, and we have

$$x'_l x'_l = c^2 t'^2. \tag{12.2.12}$$

Thus at this stage we have, in the unprimed system

$$x_1^2 + x_2^2 + x_3^2 - c^2 t^2 = 0, \tag{12.2.13}$$

and in the primed system

$$x_1'^2 + x_2'^2 + x_3'^2 - c^2 t'^2 = 0; \tag{12.2.14}$$

but from the transformation relations in (12.2.9), which have already been obtained, and (12.2.13) and (12.2.14) we have

$$x_1^2 - c^2 t^2 = x_1'^2 - c^2 t'^2. \tag{12.2.15}$$

Since, as noted earlier, no preference be given to any point over another, the transformation between coordinate systems must be linear, i.e., x'_1 and t' *must* be *linear* functions of x_1 and t *alone*. Clearly, then dropping the subscripts we may write

$$x' = \alpha x + \beta t, \quad t' = \gamma x + \delta t, \tag{12.2.16}$$

where the *constants* $\alpha, \beta, \gamma, \delta$ are to be determined so that the quadratic relation in (12.2.15) is satisfied for all x and t.

For the origin $0'$ we have $x' = 0$, hence from (12.2.16), we find

$$x = -\frac{\beta}{\alpha} t, \tag{12.2.17}$$

and, since the steady velocity of $0'$ relative to 0 is v, and $t = 0$ when $x = 0$, we have

$$\frac{x}{t} = v = -\frac{\beta}{\alpha}. \tag{12.2.18}$$

For the origin 0 we have $x = 0$, hence from (12.2.16) we find

$$x' = \beta t, \qquad t' = \delta t, \tag{12.2.19}$$

and upon elimination of t from (12.2.19) we obtain

$$x' = \frac{\beta}{\delta} t', \tag{12.2.20}$$

for the motion of 0 relative to 0′. Since by symmetry the steady velocity of 0 relative to 0′ is $-v$, and $t'=0$ when $x'=0$, we have

$$\frac{x'}{t'} = -v = \frac{\beta}{\delta}, \tag{12.2.21}$$

from which, with (12.2.18), we may conclude that

$$\delta = \alpha. \tag{12.2.22}$$

Clearly then, at this stage, the transformation equations (12.2.16) take the form

$$x' = \alpha(x - vt), \qquad t' = \gamma x + \alpha t, \tag{12.2.23}$$

and substituting from (12.2.23) into (12.2.15), we obtain

$$(1 - \alpha^2 + c^2\gamma^2)x^2 - \left(1 + \alpha^2\frac{v^2}{c^2} - \alpha^2\right)c^2t^2 + 2\alpha(\alpha v + c^2\gamma)xt = 0. \tag{12.2.24}$$

Since (12.2.24) must hold for *arbitrary* x and t, the coefficients of each term must vanish and since $\alpha \neq 0$, we have

$$\alpha^2 = \frac{1}{1-(v^2/c^2)}, \qquad \gamma^2 = \frac{\alpha^2 - 1}{c^2}, \qquad \gamma = -\frac{\alpha v}{c^2}, \tag{12.2.25}$$

which constitute three equations for the determination of the two quantities α and γ. From $(12.2.25)_1$ we have

$$\alpha = +1/\sqrt{1-(v^2/c^2)}, \tag{12.2.26}$$

the + sign is taken so that in the limit of small velocities or infinite c Eqs. (12.2.23) reduce to the ordinary Galilean transformations. From $(12.2.25)_3$ we find

$$\gamma = -v/c^2\sqrt{1-(v^2/c^2)}. \tag{12.2.27}$$

Now, we must show that $(12.2.25)_2$ is consistent with (12.2.26) and (12.2.27). This may be seen by squaring (12.2.27) to obtain

$$\gamma^2c^2 = \frac{\alpha^2v^2}{c^2}, \tag{12.2.28}$$

and squaring (12.2.26), which is then solved for α^2v^2/c^2 to obtain

$$\frac{\alpha^2v^2}{c^2} = \alpha^2 - 1, \tag{12.2.29}$$

and then noting that the substitution of (12.2.28) into (12.2.29) yields $(12.2.25)_2$.

Thus in this special case of rectilinear motion in the x_1 direction the transformation is

$$x_1' = \frac{x_1 - vt}{\sqrt{1 - v^2/c^2}}, \qquad x_2' = x_2, \qquad x_3' = x_3,$$
$$t' = \frac{t - vx_1/c^2}{\sqrt{1 - v^2/c^2}}. \tag{12.2.30}$$

These are the equations of the Lorentz transformation in the simplest special case of rectilinear motion in the x_1 direction. We will not bother to obtain the transformation equations for uniform rectilinear motion in an arbitrary direction[29] because this is as much as we need for our purposes.

Note that the transformation equations in (12.2.30) can be written

$$x_1' = \alpha(x_1 - vt), \qquad t' = \alpha\left(t - \frac{vx_1}{c^2}\right), \qquad x_2' = x_2, \qquad x_3' = x_3. \tag{12.2.31}$$

Note further that if we solve (12.2.31) for x_1 and t in terms of x_1' and t', we obtain the inverse transformation, which may be written in the form

$$x_1 = \alpha(x_1' + vt'), \qquad t = \alpha\left(t' + \frac{vx_1'}{c^2}\right), \qquad x_2 = x_2', \qquad x_3 = x_3', \tag{12.2.32}$$

as expected from our earlier discussion of the relative motion of the two frames.

The more general Lorentz transformations are more easily discussed within the framework of the 4-dimensional space-time continuum[30]. Since the more general transformations are not needed for our purposes we will not bother with this.

As we already know, the corresponding classical-Galilean-transformation equations which describe the same situation are

$$x_1' = x - vt, \qquad x_2' = x_2, \qquad x_3' = x_3, \qquad t' = t, \tag{12.2.33}$$

and according to the Lorentz transformation equations (12.2.31), these equations are a good approximation if $v \ll c$. According to the Lorentz equations, length measurements at right angles to the relative motion are the same in both inertial frames. However, length measurements in the direction of the relative motion will cause lengths to appear contracted in the direction of relative motion to the observer moving with respect to the lengths and the ratio of the contraction will be $[1 - (v^2/c^2)]^{1/2}$. Similarly, clocks which are moving with respect to the observer will appear to go slow by the same retardation factor. These contractions and retardations are mutual[31]; i.e., the situation is the same for observations from $0'$ to 0 as for 0 to $0'$.

12.3 Transformation of the Electromagnetic Field Variables Resulting from Relative Motion

We now investigate the influence of this special Lorentz transformation on Maxwell's equations (10.1.36)–(10.1.38), which in indicial notation take the form

$$e_{ijk}\frac{\partial H_k}{\partial x_j}=\frac{4\pi}{c}J_i+\frac{1}{c}\frac{\partial D_i}{\partial t}, \tag{12.3.1}$$

$$e_{ijk}\frac{\partial E_k}{\partial x_j}=-\frac{1}{c}\frac{\partial B_i}{\partial t}, \tag{12.3.2}$$

$$\frac{\partial D_k}{\partial x_k}=4\pi\rho,\quad \frac{\partial B_k}{\partial x_k}=0. \tag{12.3.3}$$

We proceed in the most straightforward way, which is the one used by Einstein in his original paper,[7] and thereby avoid consideration of four-dimensional space-time.[32] Let us confine our attention to the four equations containing **E** and **B** only, i.e., (12.3.2) and $(12.3.3)_2$, and rewrite them in terms of the primed coordinates, thus

$$e_{ijk}\left[\frac{\partial x'_l}{\partial x_k}\frac{\partial E_k}{\partial x'_l}+\frac{\partial t'}{\partial x_k}\frac{\partial E_k}{\partial t'}\right]+\frac{1}{c}\left[\frac{\partial B_i}{\partial x'_l}\frac{\partial x'_l}{\partial t}+\frac{\partial B_i}{\partial t'}\frac{\partial t'}{\partial t}\right]=0 \tag{12.3.4}$$

$$\left[\frac{\partial x'_l}{\partial x_k}\frac{\partial B_k}{\partial x'_l}+\frac{\partial t'}{\partial x_k}\frac{\partial B_k}{\partial t'}\right]=0. \tag{12.3.5}$$

The transformation derivatives may readily be obtained from (12.2.31), which yield

$$\frac{\partial x'_1}{\partial x_1}=\alpha,\quad \frac{\partial x'_1}{\partial t}=-\alpha v,\quad \frac{\partial t'}{\partial t}=\alpha,\quad \frac{\partial t'}{\partial x_1}=-\alpha\frac{v}{c^2},\quad \frac{\partial x'_2}{\partial x_2}=1,\quad \frac{\partial x'_3}{\partial x_3}=1 \tag{12.3.6}$$

and the other 10 derivatives vanish. Now, writing Eqs. $(12.3.3)_2$ and (12.3.2) out fully in terms of the unprimed coordinate system we have, respectively,

$$\frac{\partial B_1}{\partial x_1}+\frac{\partial B_2}{\partial x_2}+\frac{\partial B_3}{\partial x_3}=0, \tag{12.3.7}$$

$$\frac{\partial E_3}{\partial x_2}-\frac{\partial E_2}{\partial x_3}+\frac{1}{c}\frac{\partial B_1}{\partial t}=0,$$

$$\frac{\partial E_1}{\partial x_3}-\frac{\partial E_3}{\partial x_1}+\frac{1}{c}\frac{\partial B_2}{\partial t}=0,$$

$$\frac{\partial E_2}{\partial x_1}-\frac{\partial E_1}{\partial x_2}+\frac{1}{c}\frac{\partial B_3}{\partial t}=0. \tag{12.3.8}$$

Rewriting Eqs. (12.3.7) and (12.3.8) in terms of the primed coordinates, substituting for the transformation derivatives from (12.3.6) and rearranging terms, we obtain

$$\frac{\partial B_2}{\partial x'_2}+\frac{\partial B_3}{\partial x'_3}+\alpha\frac{\partial B_1}{\partial x'_1}-\alpha\frac{v}{c^2}\frac{\partial B_1}{\partial t'}=0 \tag{12.3.9}$$

$$\frac{\partial E_3}{\partial x'_2}-\frac{\partial E_2}{\partial x'_3}-\alpha\frac{v}{c}\frac{\partial B_1}{\partial x'_1}+\frac{\alpha}{c}\frac{\partial B_1}{\partial t'}=0, \tag{12.3.10}$$

$$-\frac{\partial}{\partial x'_1}\left[\alpha\left(E_3+\frac{v}{c}B_2\right)\right]+\frac{\partial E_1}{\partial x'_3}+\frac{1}{c}\frac{\partial}{\partial t'}\left[\alpha\left(B_2+\frac{v}{c}E_3\right)\right]=0, \tag{12.3.11}$$

$$-\frac{\partial E_1}{\partial x'_2}+\frac{\partial}{\partial x'_1}\left[\alpha\left(E_2-\frac{v}{c}B_3\right)\right]+\frac{1}{c}\frac{\partial}{\partial t'}\left[\alpha\left(B_3-\frac{v}{c}E_2\right)\right]=0. \tag{12.3.12}$$

Equations (12.3.11) and (12.3.12) are in a form just like the unprimed equations in $(12.3.8)_{2-3}$ and linear combinations of (12.3.9) and (12.3.10) can produce two equations like (12.3.7) and $(12.3.8)_1$. To see this multiply (12.3.9) by α, (12.3.10) by $\alpha v/c$ and add to obtain

$$\frac{\partial}{\partial x'_3}\left[\alpha\left(B_3-\frac{v}{c}E_2\right)\right]+\frac{\partial}{\partial x'}\left[\alpha\left(B_2+\frac{v}{c}E_3\right)\right]+\alpha^2\left(1-\frac{v^2}{c^2}\right)\frac{\partial B_1}{\partial x'_1}=0. \tag{12.3.13}$$

Now, multiply (12.3.9) by $\alpha v/c$, (12.3.10) by α and add to obtain

$$-\frac{\partial}{\partial x'_3}\left[\alpha\left(E_2-\frac{v}{c}B_3\right)\right]+\frac{\partial}{\partial x'_2}\left[\alpha\left(E_3+\frac{v}{c}B_2\right)\right]+\frac{\alpha^2}{c}\left(1-\frac{v^2}{c^2}\right)\frac{\partial B_1}{\partial t'}=0. \tag{12.3.14}$$

By virtue of $(12.2.25)_1$, Eqs. (12.3.11)–(12.3.14) are of the same form in the primed coordinates as the starting equations (12.3.7) and (12.3.8) in the unprimed system in accordance with Einstein's hypothesis of the constancy of the speed of light relative to any inertial frame, and the consequent invariance of Maxwell's equations, which describe the propagation of light, with respect to inertial frames, i.e., under inertial Lorentz transformations. Accordingly, the transformation equations for the field vectors **E** and **B** must be of the form

$$\begin{aligned}
E'_1&=E_1, & B'_1&=B_1,\\
E'_2&=\alpha\left(E_2-\frac{v}{c}B_3\right), & B'_2&=\alpha\left(B_2+\frac{v}{c}E_3\right),\\
E'_3&=\alpha\left(E_3+\frac{v}{c}B_2\right), & B'_3&=\alpha\left(B_3-\frac{v}{c}E_2\right).
\end{aligned} \tag{12.3.15}$$

Clearly, with the transformations of the components of the field vectors in (12.3.15), the four equations referred to the primed inertial system may be written

$$\frac{\partial E'_1}{\partial x'_3} - \frac{\partial E'_3}{\partial x'_1} + \frac{1}{c}\frac{\partial B'_2}{\partial t'} = 0,$$

$$\frac{\partial E'_2}{\partial x'_1} - \frac{\partial E'_1}{\partial x'_2} + \frac{1}{c}\frac{\partial B'_3}{\partial t'} = 0,$$

$$\frac{\partial E'_3}{\partial x'_2} - \frac{\partial E'_2}{\partial x'_2} + \frac{1}{c}\frac{\partial B'_1}{\partial t'} = 0, \quad (12.3.16)$$

$$\frac{\partial B'_1}{\partial x'_1} + \frac{\partial B'_2}{\partial x'_2} + \frac{\partial B'_3}{\partial x'_3} = 0. \quad (12.3.17)$$

Equations (12.3.16) and (12.3.17) show explicitly that with the transformations of **E** and **B** in (12.3.15), Maxwell's Faraday equations and the scalar equation on **B** indeed take the same form in any inertial system. Since the direction of linear translation could, at most, be a single arbitrary direction with respect to the Cartesian axes, from our results in (12.3.15) we can write the Lorentz transformation of the field vectors in the more general form

$$\mathbf{E}'_p = \mathbf{E}_p, \qquad \mathbf{B}'_p = \mathbf{B}_p,$$

$$\mathbf{E}'_n = \alpha\left(\mathbf{E}_n + \frac{1}{c}\mathbf{v} \times \mathbf{B}_n\right), \quad \mathbf{B}'_n = \alpha\left(\mathbf{B}_n - \frac{1}{c}\mathbf{v} \times \mathbf{E}_n\right), \quad (12.3.18)$$

where the subscripts p and n, respectively, stand for parallel and normal to $\mathbf{v}$. The transformation equations in (12.3.18) are in Guassian units. Since $\mathbf{v} \times \mathbf{E}_p = \mathbf{v} \times \mathbf{B}_p = 0$, Eqs. (12.3.18) can also be written

$$\mathbf{E}'_p = \left(\mathbf{E}_p + \frac{1}{c}\mathbf{v} \times \mathbf{B}_p\right), \quad \mathbf{B}'_p = \left(\mathbf{B}_p - \frac{1}{c}\mathbf{v} \times \mathbf{E}_p\right),$$

$$\mathbf{E}'_n = \alpha\left(\mathbf{E}_n + \frac{1}{c}\mathbf{v} \times \mathbf{B}_n\right), \quad \mathbf{B}'_n = \alpha\left(\mathbf{B}_n - \frac{1}{c}\mathbf{v} \times \mathbf{E}_n\right). \quad (12.3.19)$$

Let us now confine our attention to the four equations containing **H**, **D**, ρ and **J** only, i.e., (12.3.1) and $(12.3.3)_1$ and carry out the transformation from the unprimed to the primed inertial system. When written out fully Eqs. $(12.3.3)_1$ and (12.3.1) in the unprimed coordinate system take the respective forms

$$\frac{\partial D_1}{\partial x_1} + \frac{\partial D_2}{\partial x_2} + \frac{\partial D_3}{\partial x_3} = 4\pi\rho, \quad (12.3.20)$$

$$\frac{\partial H_3}{\partial x_2} - \frac{\partial H_2}{\partial x_3} - \frac{4\pi}{c}J_1 - \frac{1}{c}\frac{\partial D_1}{\partial t} = 0,$$

$$\frac{\partial H_1}{\partial x_3} - \frac{\partial H_3}{\partial x_1} - \frac{4\pi}{c}J_2 - \frac{1}{c}\frac{\partial D_2}{\partial t} = 0,$$

$$\frac{\partial H_2}{\partial x_1} - \frac{\partial H_1}{\partial x_2} - \frac{4\pi}{c}J_3 - \frac{1}{c}\frac{\partial D_3}{\partial t} = 0. \quad (12.3.21)$$

Rewriting Eqs. (12.3.20) and (12.3.21) in terms of the primed coordinates, substituting for the transformation derivatives from (12.3.6) and rearranging terms, we obtain

$$\alpha\frac{\partial D_1}{\partial x_1'} - \frac{\alpha v}{c^2}\frac{\partial D_1}{\partial t'} + \frac{\partial D_2}{\partial x_2'} + \frac{\partial D_3}{\partial x_3'} - 4\pi\rho = 0, \tag{12.3.22}$$

$$\frac{\partial H_3}{\partial x_2'} - \frac{\partial H_2}{\partial x_3'} + \frac{\alpha v}{c}\frac{\partial D_1}{\partial x_1'} - \frac{\alpha}{c}\frac{\partial D_1}{\partial t'} - \frac{4\pi}{c}J_1 = 0, \tag{12.3.23}$$

$$\frac{\partial H_1}{\partial x_3'} - \frac{\partial}{\partial x_1'}\left[\alpha\left(H_3 - \frac{v}{c}D_2\right)\right] - \frac{4\pi}{c}J_2 - \frac{1}{c}\frac{\partial}{\partial t'}\left[\alpha\left(D_2 - \frac{v}{c}H_3\right)\right] = 0, \tag{12.3.24}$$

$$\frac{\partial}{\partial x_1'}\left[\alpha\left(H_2 + \frac{v}{c}D_3\right)\right] - \frac{\partial H_1}{\partial x_2'} - \frac{4\pi}{c}J_3 - \frac{1}{c}\frac{\partial}{\partial t'}\left[\alpha\left(D_3 + \frac{v}{c}H_2\right)\right] = 0. \tag{12.3.25}$$

Equations (12.3.24) and (12.3.25) are in a form just like the unprimed equations in $(12.3.21)_{2-3}$ and linear combinations of (12.3.22) and (12.3.23) can produce two equations like (12.3.20) and $(12.3.21)_1$. To see this multiply (12.3.22) by α and (12.3.23) by $\alpha v/c$ and subtract the second from the first to obtain

$$\frac{\partial}{\partial x_2'}\left[\alpha\left(D_2 - \frac{v}{c}H_3\right)\right] + \frac{\partial}{\partial x_3'}\left[\alpha\left(D_3 + \frac{v}{c}H_2\right)\right] + \alpha^2\left(1 - \frac{v^2}{c^2}\right)\frac{\partial D_1}{\partial x_1'} - 4\pi\left[\alpha\left(\rho - \frac{vJ_1}{c^2}\right)\right] = 0. \tag{12.3.26}$$

Now, multiply (12.3.22) by $\alpha v/c$ and (12.3.23) by α and subtract the first from the second to obtain

$$\frac{\partial}{\partial x_2'}\left[\alpha\left(H_3 - \frac{v}{c}D_2\right)\right] - \frac{\partial}{\partial x_3'}\left[\alpha\left(H_2 + \frac{v}{c}D_3\right)\right] - \frac{4\pi}{c}[\alpha(J_1 - \rho v)] - \frac{1}{c}(\alpha^2)\left(1 - \frac{v^2}{c^2}\right)\frac{\partial D_1}{\partial t'} = 0. \tag{12.3.27}$$

By virtue of $(12.2.25)_1$, Eqs. (12.3.24)–(12.3.27) in the primed inertial frame are in exactly the same form as the starting equations (12.3.20) and (12.3.21) in the unprimed inertial frame as required by Einstein's hypothesis. We have only to write the transformations of the field vectors **D** and **H** in the form

$$\begin{aligned} D_1' &= D_1, & H_1' &= H_1, \\ D_2' &= \alpha\left(D_2 - \frac{v}{c}H_3\right), & H_2' &= \alpha\left(H_2 + \frac{v}{c}D_3\right), \\ D_3' &= \alpha\left(D_3 + \frac{v}{c}H_2\right), & H_3' &= \alpha\left(H_3 - \frac{v}{c}D_2\right), \end{aligned} \tag{12.3.28}$$

and the transformations of the charge and current densities in the form

$$\rho' = \alpha\left(\rho - \frac{vJ_1}{c^2}\right), \quad J'_1 = \alpha(J_1 - \rho v),$$

$$J'_2 = J_2, \qquad J'_3 = J_3. \tag{12.3.29}$$

Clearly, with the transformations in (12.3.28) and (12.3.29), the four equations referred to the primed inertial system may be written

$$\frac{\partial H'_1}{\partial x'_3} - \frac{\partial H'_3}{\partial x'_1} - \frac{4\pi}{c}J'_2 - \frac{1}{c}\frac{\partial D'_2}{\partial t'} = 0,$$

$$\frac{\partial H'_2}{\partial x'_1} - \frac{\partial H'_1}{\partial x'_2} - \frac{4\pi}{c}J'_3 - \frac{1}{c}\frac{\partial D'_3}{\partial t'} = 0,$$

$$\frac{\partial H'_3}{\partial x'_2} - \frac{\partial H'_2}{\partial x'_3} - \frac{4\pi}{c}J'_1 - \frac{1}{c}\frac{\partial D'_1}{\partial t'} = 0, \tag{12.3.30}$$

$$\frac{\partial D'_1}{\partial x'_1} + \frac{\partial D'_2}{\partial x'_2} + \frac{\partial D'_3}{\partial x'_3} - 4\pi\rho' = 0. \tag{12.3.31}$$

Equations (12.3.30) and (12.3.31) show explicitly that with the transformations of $\mathbf{H}, \mathbf{D}, \rho$, and $\mathbf{J}$ in (12.3.28) and (12.3.29), Maxwell's corrected version of Ampere's equations and the charge equation of electrostatics indeed take the same form in any inertial system. Again, considering an arbitrary direction of linear translation, from the results in (12.3.28) and (12.3.29) we can write the Lorentz transformation of these fields in the more general form

$$\mathbf{D}'_p = \mathbf{D}_p, \qquad \mathbf{H}'_p = \mathbf{H}_p,$$

$$\mathbf{D}'_n = \alpha\left(\mathbf{D}_n + \frac{1}{c}\mathbf{v} \times \mathbf{H}_n\right), \quad \mathbf{H}'_n = \alpha\left(\mathbf{H}_n - \frac{1}{c}\mathbf{v} \times \mathbf{D}_n\right), \tag{12.3.32}$$

$$\rho' = \alpha\left(\rho - \frac{\mathbf{v}\cdot\mathbf{J}}{c^2}\right), \quad \mathbf{J}'_p = \alpha(\mathbf{J}_p - \rho\mathbf{v}), \quad \mathbf{J}'_n = \mathbf{J}_n, \tag{12.3.33}$$

and, as in the case of **E** and **B** before, and for the same reason, the parallel field-transformations in Eqs. $(12.3.32)_{1-2}$ may be written

$$\mathbf{D}'_p = \left(\mathbf{D} + \frac{1}{c}\mathbf{v} \times \mathbf{H}_p\right), \quad \mathbf{H}'_p = \left(\mathbf{H} - \frac{1}{c}\mathbf{v} \times \mathbf{D}_p\right). \tag{12.3.34}$$

As stated previously, although we are interested in the transformations of the field vectors **E** and **B**, which belong to the space, and the hybrids **D** and **H**, for our present purposes we are primarily interested in the transformations of the material vectors **P** and **M**, which are composed of the differences of the field vectors **E** and **B** and the hybrids **D** and **H**. We are also very interested in the transformations of the material quantities ρ and **J**. Now physically, the principle of invariance clearly says that not only is the form taken by Maxwell's equations in every inertial system the same, but all other nonspecializing, intrinsic, general relations between the field vectors must be preserved, since there is nothing to distinguish any one special inertial system from the rest in any absolute sense. In particular, since we have

$$\mathbf{D} = \mathbf{E} + 4\pi\mathbf{P}, \quad \mathbf{H} = \mathbf{B} - 4\pi\mathbf{M}, \tag{12.3.35}$$

we must also have

$$\mathbf{D}' = \mathbf{E}' + 4\pi\mathbf{P}', \quad \mathbf{H}' = \mathbf{B}' - 4\pi\mathbf{M}'. \tag{12.3.36}$$

From our transformation relations (12.3.19), $(12.3.32)_{3-4}$ and (12.3.34) and (12.3.35) and (12.3.36), we can readily obtain the transformation relations for **P** and **M** in the form

$$\mathbf{P}'_p = \left[\mathbf{P}_p - \frac{1}{c}\mathbf{v} \times \mathbf{M}_p\right], \quad \mathbf{M}'_p = \left[\mathbf{M}_p + \frac{1}{c}\mathbf{v} \times \mathbf{P}_p\right],$$

$$\mathbf{P}'_n = \alpha\left[\mathbf{P}_n - \frac{1}{c}\mathbf{v} \times \mathbf{M}_n\right], \quad \mathbf{M}'_n = \alpha\left[\mathbf{M}_n + \frac{1}{c}\mathbf{v} \times \mathbf{P}_n\right]. \tag{12.3.37}$$

Equations (12.3.37) are the transformation equations of the material vectors **P** and **M**, which are of primary interest to us here along with the transformation relations in (12.3.33) for ρ and **J**. Our interest in the transformation relations in (12.3.19), (12.3.33) and (12.3.37) is the primary reason we developed the special theory of relativity. We are particularly interested in these transformation relations when the relative speeds of the two inertial frames $|\mathbf{v}| \ll c$, i.e., in the nonrelativistic limit. Under these circumstances we may neglect v^2/c^2 while still retaining $\mathbf{v}/c$. The transformation equations in (12.3.19), (12.3.33) and (12.3.37), respectively, of the space and matter fields of interest then take the simpler forms

$$\mathbf{E}' = \mathbf{E} + \frac{1}{c}\mathbf{v} \times \mathbf{B}, \quad \mathbf{B}' = \mathbf{B} - \frac{1}{c}\mathbf{v} \times \mathbf{E}, \tag{12.3.38}$$

$$\rho' = \rho, \quad \mathbf{J}' = \mathbf{J} - \rho\mathbf{v}, \tag{12.3.39}$$

$$\mathbf{P}' = \mathbf{P} - \frac{1}{c}\mathbf{v} \times \mathbf{M}, \quad \mathbf{M}' = \mathbf{M} + \frac{1}{c}\mathbf{v} \times \mathbf{P}. \tag{12.3.40}$$

In the low-velocity transformation equations (12.3.38)–(12.3.40) the primed variables are referred to the system moving with velocity **v** relative to the stationary system, to which the unprimed variables are referred. Although the transformation relations in (12.3.38) for the fields **E** and **B**, which belong to the space, are in the form of direct interest in the description of deformable continua, there are times when the transformation relations inverse to (12.3.39) and (12.3.40) are required for the fields ρ, **J**, **P** and **M**, which belong to the matter. Since in the low velocity limit of the relativistic transformation equations terms in v^2/c^2 are negligible, it is clear that the inverse relations are given by[33]

$$\mathbf{E} = \mathbf{E}' - \frac{1}{c}\mathbf{v} \times \mathbf{B}', \quad \mathbf{B} = \mathbf{B}' + \frac{1}{c}\mathbf{v} \times \mathbf{E}', \tag{12.3.41}$$

$$\rho = \rho', \quad \mathbf{J} = \mathbf{J}' + \rho\mathbf{v}, \tag{12.3.42}$$

$$\mathbf{P} = \mathbf{P}' + \frac{1}{c}\mathbf{v} \times \mathbf{M}', \quad \mathbf{M} = \mathbf{M}' - \frac{1}{c}\mathbf{v} \times \mathbf{P}'. \tag{12.3.43}$$

It should be noted that although Eqs. (12.3.38)–(12.3.43) were obtained by transforming from one uniform inertial coordinate system to another, we are implicitly indicating that they can be applied in the case of a deforming and accelerating continuum, in which each point is moving not only at a velocity $\mathbf{v}$ which is changing in time but in space as well. This is physically sensible from the point of view that the nonrelativistic limit of the transformation equations of electromagnetic quantities is valid for small velocities, velocities sufficiently small that velocity, acceleration, position and time can very accurately be regarded as Galilean and nonrelativistic. This is clearly a very accurate approximation for the description of a deformable continuum. Under this approximation, it should be clear that we are effectively assuming that the nonrelativistic limit of the transformations of electromagnetic quantities are the *approximate* Galilean transformations of the electromagnetic quantities. This is clearly physically reasonable since it is what is done in reducing *Lorentz* position and velocity transformations to Galilean transformations. This approximation is also made in microscopic solid-state physics[34] where the particles are moving with *small* velocities compared with the speed of light. However, it should be noted that although the position, velocity and time can be treated in an *exactly* Galilean invariant manner, the electromagnetic field variables can be treated only in an *approximately* Galilean invariant manner because the Maxwell field equations (10.1.36)–(10.1.38) are only Lorentz invariant.

CHAPTER 13

The Electromagnetic Potentials

13.1 The Electromagnetic Potential Formulation of Maxwell's Equations

The potential formulation of the electromagnetic field equations is very useful and has many important applications. The particular applications of interest to us are in the description of the quasi-static electric and magnetic fields. It should be noted that in this formulation we are concerned with the field equations and Poynting's theorem only, and not in the equations that involve force and momentum.

We begin with Maxwell's equations (10.1.36) and (10.1.37), which take the form

$$\nabla \times \mathbf{H} = \frac{1}{c}\frac{\partial \mathbf{D}}{\partial t} + \frac{4\pi}{c}\mathbf{J}, \tag{13.1.1}$$

$$\nabla \times \mathbf{E} = -\frac{1}{c}\frac{\partial \mathbf{B}}{\partial t}. \tag{13.1.2}$$

We also have the relations in (10.1.39), which we rewrite here thus

$$\mathbf{D} = \mathbf{E} + 4\pi\mathbf{P}, \quad \mathbf{H} = \mathbf{B} - 4\pi\mathbf{M}, \tag{13.1.3}$$

and the auxiliary equations in (10.1.38), which are

$$\nabla \cdot \mathbf{B} = 0, \quad \nabla \cdot \mathbf{D} = 4\pi\rho, \tag{13.1.4}$$

the time derivative of the first of which is satisfied automatically, and the second of which simply defines the electric charge density ρ such that the equation of the conservation of electric charge (10.1.28), i.e.,

$$\partial\rho/\partial t + \nabla \cdot \mathbf{J} = 0 \tag{13.1.5}$$

is satisfied. In addition to these equations, we also have Poynting's theorem (11.1.6), which we rewrite here in the form

$$-\frac{c}{4\pi}\int_S \mathbf{n} \cdot \mathbf{E} \times \mathbf{H}\, dS = \frac{1}{4\pi}\int_V \left(\mathbf{E} \cdot \frac{\partial \mathbf{D}}{\partial t} + \mathbf{H} \cdot \frac{\partial \mathbf{B}}{\partial t}\right) dV + \int_V \mathbf{E} \cdot \mathbf{J}\, dV. \tag{13.1.6}$$

We now note that Maxwell's equations can be reformulated in terms of the vector and scalar potentials **A** and φ. Accordingly, we first observe that as a result of $(13.1.4)_1$ we have

$$\mathbf{B} = \nabla \times \mathbf{A}, \tag{13.1.7}$$

but although **B** is unique, **A** is not since we have

$$\mathbf{B} = \nabla \times \bar{\mathbf{A}}, \tag{13.1.8}$$

also, where

$$\bar{\mathbf{A}} = \mathbf{A} + \nabla\psi, \tag{13.1.9}$$

and ψ is any scalar, because

$$\nabla \times \nabla\psi \equiv 0. \tag{13.1.10}$$

Note that $\nabla \cdot \mathbf{A}$ is arbitrary for a unique **B**. Different selections of $\nabla \cdot \mathbf{A}$ determine what are called different electromagnetic gauges, and once a selection of $\nabla \cdot \mathbf{A}$ has been made, **A** is unique. For example, if we set $\nabla \cdot \mathbf{A} = 0$, **A** may be determined uniquely if **B** is known. The substitution of (13.1.7) into (13.1.2) yields

$$\nabla \times \left(\mathbf{E} + \frac{1}{c}\frac{\partial \mathbf{A}}{\partial t}\right) = 0, \tag{13.1.11}$$

which enables us to write

$$\mathbf{E} = -\nabla\varphi - \frac{1}{c}\frac{\partial \mathbf{A}}{\partial t}, \tag{13.1.12}$$

and clearly φ is not unique since **A** is not unique. However, by direct substitution of (13.1.9) in (13.1.12) it is clear that we have

$$\mathbf{E} = -\nabla\bar{\varphi} - \frac{1}{c}\frac{\partial \bar{\mathbf{A}}}{\partial t}, \tag{13.1.13}$$

where

$$\bar{\varphi} = \varphi - \frac{1}{c}\frac{\partial \psi}{\partial t}. \tag{13.1.14}$$

Thus, in terms of the (non-unique) vector and scalar potentials, Maxwell's equations (13.1.1). (13.1.2) and (13.1.4), may be written in the form

$$\nabla \times \mathbf{H} = \frac{1}{c}\frac{\partial \mathbf{D}}{\partial t} + \frac{4\pi}{c}\mathbf{J}, \tag{13.1.15}$$

$$\mathbf{B} = \nabla \times \mathbf{A}, \tag{13.1.16}$$

$$\mathbf{E} = -\nabla\varphi - \frac{1}{c}\frac{\partial \mathbf{A}}{\partial t}, \tag{13.1.17}$$

and we still have (13.1.3), i.e.,

$$\mathbf{D} = \mathbf{E} + 4\pi\mathbf{P}, \quad \mathbf{H} = \mathbf{B} - 4\pi\mathbf{M}, \tag{13.1.18}$$

along with the auxiliary relation $(13.1.4)_2$, i.e.,

$$\nabla \cdot \mathbf{D} = 4\pi\rho. \tag{13.1.19}$$

Now, the electromagnetic potentials $\mathbf{A}$ and φ enable us to put Poynting's theorem (13.1.6) in a form which is particularly useful for making quasi-static approximations. Consider the term

$$\nabla \cdot \mathbf{h} = \nabla \cdot \left(\frac{c}{4\pi} \mathbf{E} \times \mathbf{H} \right), \tag{13.1.20}$$

in Poynting's differential equation (11.1.5), which is responsible for the Poynting energy flux

$$\int_S \mathbf{n} \cdot \mathbf{h} \, dS = \int_S \mathbf{n} \cdot \left(\frac{c}{4\pi} \mathbf{E} \times \mathbf{H} \right) dS, \tag{13.1.21}$$

in the integral form of Poynting's equation (13.1.6). In indicial notation Eq. (13.1.20) may be written

$$h_{i,i} = \frac{c}{4\pi} [e_{ijk} E_j H_k]_{,i}, \tag{13.1.22}$$

and substituting for $\mathbf{E}$ from (13.1.17), we obtain

$$h_{i,i} = -\frac{c}{4\pi} \left[e_{ijk} \left(\varphi_{,j} + \frac{1}{c} \dot{A}_j \right) H_k \right]_{,i} = -\frac{c}{4\pi} e_{ijk} \left[\varphi_{,j} H_{k,i} + \frac{1}{c} (\dot{A}_j H_k)_{,i} \right], \tag{13.1.23}$$

and the substitution of (13.1.15) into (13.1.23) yields

$$\begin{aligned} h_{i,i} &= \frac{c}{4\pi} \varphi_{,j} \left(\frac{1}{c} \dot{D}_j + \frac{4\pi}{c} J_j \right) - \frac{1}{4\pi} (e_{ijk} \dot{A}_j H_k)_{,i} \\ &= \frac{1}{4\pi} [(\varphi \dot{D}_j)_{,j} - \varphi \dot{D}_{j,j}] + (\varphi J_j)_{,j} - \varphi J_{j,j} - \frac{1}{4\pi} (e_{ijk} \dot{A}_j H_k)_{,i}, \end{aligned}$$

which, with the divergence of (13.1.15) enables us to write

$$h_{i,i} = \left[\varphi \left(\frac{\dot{D}_i}{4\pi} + J_i \right) - \frac{1}{4\pi} e_{ijk} \dot{A}_j H_k \right]_{,i}. \tag{13.1.24}$$

Substituting from this equation into Poynting's differential equation (11.1.5), inegrating over an arbitrary volume and employing the divergence theorem, we get Poynting's theorem in the form

$$-\int_S n_i \left[\varphi \left(\frac{\dot{D}_i}{4\pi} + J_i \right) - \frac{1}{4\pi} e_{ijk} \dot{A}_j H_k \right] dS = \int_V \frac{1}{4\pi} (E_i \dot{D}_i + H_i \dot{B}_i) \, dV + \int_V E_i J_i \, dV, \tag{13.1.25}$$

and, as a consequence, we obtain

$$h_i = \varphi \left(\frac{\dot{D}_i}{4\pi} + J_i \right) - \frac{1}{4\pi} e_{ijk} \dot{A}_j H_k, \tag{13.1.26}$$

as the expression for the Poynting vector $\mathbf{h}$ in terms of the vector and scalar potentials $\mathbf{A}$ and φ.

Now, whereas $h_i = (c/4\pi)e_{ijk}E_jH_k$ is unique, the present expression for h_i is not, because A_j and φ are not unique. Nevertheless, in the present case we know that $\int_S n_i h_i \, dS$, where S is any closed surface, is unique, and that is all that counts. Moreover, the unique form of $\mathbf{h}$ can have added to it $\nabla \times \mathbf{V}$, where $\mathbf{V}$ is any vector, without violating Poynting's theorem, because

$$\int_S \mathbf{n}\cdot\nabla \times \mathbf{V}\, dS = 0,$$

where S is any closed surface. That is to say, the non-unique $\mathbf{h}$'s all differ by the curl of some vector, and that does not matter because the surface integral of the curl of a vector over a closed surface vanishes. In other words, if $\bar{\mathbf{h}} \neq \mathbf{h} \neq (c/4\pi)\mathbf{E} \times \mathbf{H}$, we still have

$$\int_S \mathbf{n}\cdot\bar{\mathbf{h}}\, dS = \int_S \mathbf{n}\cdot\mathbf{h}\, dS = \int_S \mathbf{n}\cdot\frac{c}{4\pi}\mathbf{E} \times \mathbf{H}\, dS. \tag{13.1.27}$$

13.2 The Quasi-Static Electric Field

We are especially interested in the electromagnetic potential formulation in the case of polarizable (but not magnetizable) dielectrics, i.e., when[35]

$$\rho = 0, \quad J_i = 0, \quad M_i = 0, \tag{13.2.1}$$

and under these circumstances from $(13.1.18)_2$ we have

$$H_i \equiv B_i, \tag{13.2.2}$$

and H_i (and consequently A_i) cannot be zero because from (13.1.15)

$$\nabla \times \mathbf{H} = \frac{1}{c}\dot{\mathbf{D}}, \tag{13.2.2}$$

and

$$\dot{\mathbf{D}} \neq 0, \tag{13.2.4}$$

under time-dependent electromagnetic processes in the polarizable dielectric. Under these circumstances Maxwell's equations (13.1.15)–(13.1.17) in indicial notation take the form

$$e_{ijk}H_{k,j} = \frac{1}{c}\dot{D}_i, \tag{13.2.5}$$

$$H_k = e_{klm}A_{m,l}, \tag{13.2.6}$$

$$E_i = -\varphi_{,i} - \frac{1}{c}\dot{A}_i, \tag{13.2.7}$$

and from $(13.1.18)_1$ we have

$$D_i = E_i + 4\pi P_i, \tag{13.2.8}$$

and from (13.1.19) and (13.2.1) we have the auxiliary equation

$$D_{i,i} = 0. \tag{13.2.9}$$

We use this potential formulation to obtain the equations of the quasi-state electric field for the polarizable dielectric, in which electromagnetic propagation has been excluded. The basic simplifying assumption which is used to get the quasi-static equations is that each component of $\varphi_{,i}$ that enters in a problem satisfies[36]

$$|\dot{A}_i/c| \ll |\varphi_{,i}|. \tag{13.2.10}$$

This assumption is well borne out experimentally in many cases. It basically is valid when the frequency of a disturbance is sufficiently low that the wavelength of the associated electromagnetic wave is much larger than any typical length associated with the spatial rate of variation of φ.

Since we are considering a disturbance at frequency ω, we note that the quantity ω/c which arises in (13.2.7) on account of the $\dot{\mathbf{A}}/c$ term has the units of reciprocal length $(1/L)$, where L is of the order of the very large electromagnetic wavelength mentioned above. We further note that φ varies sufficiently rapidly in space that the $\nabla\varphi$ term appearing in (13.2.7) results in a reciprocal length $(1/l)$ such that $l \ll L$. It may readily be seen that $(l/L) \ll 1$ is the condition for the validity of the assumption in (13.2.10) by considering a solution of the system in the form of a successive approximation about $\dot{\mathbf{A}}/c = 0$. When $\dot{A}_i/c$ is neglected in (13.2.7), the auxiliary equation (13.2.9) may be used to find φ. When φ is thus determined, A_i may be found by means of (13.2.5) and (13.2.6) with the now known $\dot{D}_i$ from (13.2.8) on the right-hand side of (13.2.5) with the result that each component of A_i is of the order of $(l/L)\varphi$, since each gradient results in a reciprocal length. This shows that the assumption in (13.2.10) is indeed statisfied whenever $(l/L) \ll 1$. Furthermore, as a result of this the second term on the right-hand side of (13.2.7) is of the order of $(l/L)^2$ of the first term. We could now continue with this successive approximation procedure with the now known nonzero A_i and determine a second approximation for φ, and so on. However, l/L is so small that the first approximation, i.e., the theory of the quasi-static electric field, is as accurate as necessary. Moreover, when $(l/L) \ll 1$ it is clear from (13.1.26) that the magnetic portion of the Poynting energy flux is negligible [down by a factor of $(l/L)^2$] compared to the electric portion. Hence, the degenerate form of the Poynting energy flux in (13.1.26) may be written in the form

$$h_i = \varphi \dot{D}_i/4\pi, \tag{13.2.11}$$

since $J_i = 0$. In the same way it is clear that the magnetic term $H_k\dot{B}_k$ is negligible cmpared with the electric term $E_k\dot{D}_k$ in (13.1.25).

13.3 The Quasi-Static Magnetic Field for Insulators

At this point it should be clear that if we are concerned with a magnetizable dielectric, e.g., a ferrite, in which[37]

$$\rho = 0, \quad J_i = 0, \quad P_i = 0, \tag{13.3.1}$$

Maxwell's equations (13.1.1) and (13.1.2) may be written in the form

$$e_{ijk} H_{k,j} = \frac{1}{c}\dot{E}_i, \tag{13.3.2}$$

$$e_{ijk} E_{k,j} = -\frac{1}{c}\dot{B}_i, \tag{13.3.3}$$

since (13.1.3) take the form

$$D_i = E_i, \quad H_i = B_i - 4\pi M_i, \tag{13.3.4}$$

and where the auxiliary equations in (13.1.4) take the form

$$E_{i,i} = 0, \quad B_{i,i} = 0. \tag{13.3.5}$$

The formal similarity between these equations and those of the polarizable dielectric discussed in Sec. 13.2 is quite obvious. Hence, it is clear that we can define a magnetic potential system consisting of an axial vector potential $\mathbf{A}^M$ such that

$$\mathbf{E} = \nabla \times \mathbf{A}^M, \tag{13.3.6}$$

and an axial scalar φ^M such that

$$\mathbf{H} = -\nabla\varphi^M + \frac{1}{c}\dot{\mathbf{A}}^M. \tag{13.3.7}$$

At this stage it is clear that the magnetic potential formulation of Eqs. (13.3.2)–(13.3.5) may be written

$$\mathbf{E} = \nabla \times \mathbf{A}^M, \tag{13.3.8}$$

$$\mathbf{H} = -\nabla\varphi^M + \frac{1}{c}\dot{\mathbf{A}}^M, \tag{13.3.9}$$

$$\nabla \times \mathbf{E} = -\frac{1}{c}\dot{\mathbf{B}}, \tag{13.3.10}$$

along with the relation

$$\mathbf{H} = \mathbf{B} - 4\pi\mathbf{M}, \tag{13.3.11}$$

and the auxiliary equation

$$\nabla\cdot\mathbf{B} = 0. \tag{13.3.12}$$

Thus, it is clear that we may obtain the equations of the quasi-static magnetic field in exactly the same way that we obtained the equations of the quasi-static electric field and the conditions for the validity of the magnetic approximation will again be that $(l/L) \ll 1$, i.e., exactly the same as in the quasi-static electric approximation. Clearly the degenerate form of the Poynting vector in the quasi-static magnetic approximation is

$$h^M = \frac{1}{4\pi}\varphi^M \dot{\mathbf{B}}. \qquad (13.3.13)$$

Of course, in this case the electric term $E_i \dot{E}_k$ is negligible compared with the magnetic term $H_k \dot{B}_k$ in (13.1.25).

CHAPTER 14

Linear Circuit Equations from Maxwell's Equations

14.1 Electric Circuit Equations and Voltage Generation

In an elementary circuit with inductance, resistance, capacitance and applied voltage in series as shown in Figure 37 the Kirchhoff circuit equation is

$$L\ddot{I} + R\dot{I} + \frac{I}{C} = \frac{d\mathscr{E}^e}{dt}, \tag{14.1.1}$$

or

$$L\dot{I} + RI + \frac{Q}{C} = \mathscr{E}^e, \tag{14.1.2}$$

where $\mathscr{E}^e$ is the applied electromotive force or EMF, I is the current, Q is the charge on the capacitor plates, C is the capacitance, L is the inductance and R is the resistance. For two or more circuits inductively coupled the equations are

$$L_{11}\dot{I}^1 + L_{12}\dot{I}^2 + R^1 I^1 + \frac{Q^1}{C_1} = \mathscr{E}^1, \tag{14.1.3}$$

$$L_{12}\dot{I}^1 + L_{22}\dot{I}^2 + R_2 I^2 + \frac{Q^2}{C_2} = \mathscr{E}^2, \tag{14.1.4}$$

where L_{12} is the mutual inductance and L_{11} and L_{22} are the self inductances. The question naturally arises as to just what is the relation of these equations to the more general Maxwell electromagnetic field equations which must govern the circuit behavior.

As one might expect the circuit equations may be obtained by integrating Maxwell's equations with respect to space under suitable assumptions. However, as usual with approximations of this kind it is not quite as straightforward as one might expect. Indeed, before we can properly discuss this question we must understand the significance and nature of an applied EMF.

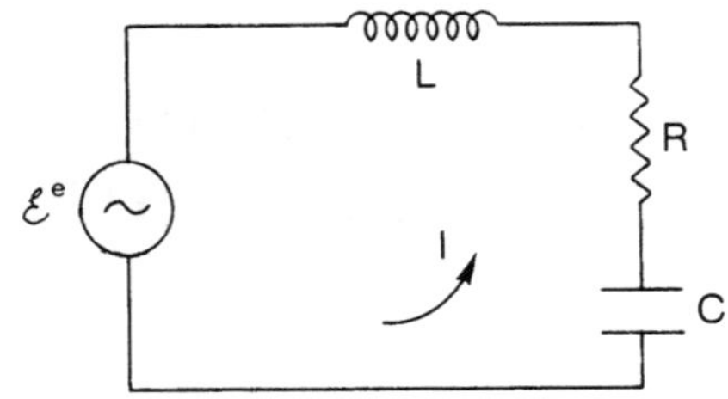

FIGURE 37. Series circuit.

An applied EMF (or voltage) does not presently exist in our differential equations explicitly, but there is nothing which prohibits us from putting it in the integral form of our equations. In particular an impressed (or applied) EMF means that there is some physical mechanism which occurs in a wet or dry cell battery, and causes the integral

$$\int_1^2 \mathbf{E}\cdot d\mathbf{r} \tag{14.1.5}$$

across the region occupied by the EMF source to be nonzero and in fact to be negative. The cause of this is an inhomogeneity in something such as, e.g., the concentration of constituents in an electrolyte solution. We are not particularly concerned with the details of this process here but with the fact that across the source we have

$$\int_1^2 \mathbf{E}\cdot d\mathbf{r} = -\mathscr{E}, \tag{14.1.6}$$

where $\mathscr{E}$ is the EMF generated by the voltage source.

14.2 Capacitance and Resistance

Since in obtaining the linear circuit equations from Maxwell's equations we will integrate Faraday's law (10.1.7) around the centerline of the circuits, we will be concerned with $\int \mathbf{E}\cdot d\mathbf{r}$ across various regions of the circuits. In addition to the integral across the voltage source, which we have already considered, we must evaluate that integral across a capacitor and a resistor. However, before we can do this we must recognize the fact that in all parts of each circuit with the exception of the capacitors, Maxwell's equations excluding the displacement current $\dot{\mathbf{D}}$ are valid because the frequency is sufficiently low for the size of the circuit that $|\dot{\mathbf{D}}| \ll |\mathbf{J}|$. This means that electromagnetic propagation may be ignored in the circuit. In other words, except for the capacitor, Ampere's law and Faraday's law are valid for the circuit. Maxwell's correction to Ampere's law is unnecessary. On the other hand, within the capacitor the frequency is low enough that the quasi-static electric field approximation, which has been developed in Sec. 13.2, holds. In other words within the capacitor the equations of electrostatics (13.2.7)–(13.2.9) without $\dot{\mathbf{A}}$ hold and are time-dependent without electromagnetic propagation.

We now consider the well-known, one-dimensional solution for the parallel plate capacitor shown in Figure 38 in which the edges are left out of account and the surface area of the electrodes is A. This is accurate enough for most practical cases. However, the real solution should consider all boundary conditions including the edges of the plate and all regions to infinity. Complex variable techniques can be used to consider two-dimensional solutions. Except for certain very simple geometries, all three-dimensional problems require some form of approximation procedure. From (13.2.9) we see that the differential equation takes the form

$$\partial D_2/\partial x_2 = 0, \tag{14.2.1}$$

and from (3.5.20) the boundary conditions are given by

$$\varphi = \pm\varphi^0 \quad \text{at} \quad x_2 = \pm h, \tag{14.2.2}$$

where φ^0 is prescribed. From (3.4.6) we also have the linear constitutive equation

$$D_2 = \varepsilon E_2 = -\varepsilon\varphi_{,2}. \tag{14.2.3}$$

The substitution of (14.2.3) into (14.2.1) yields the differential equation

$$\varphi_{,22} = 0, \tag{14.2.4}$$

which has the solution

$$\varphi = Bx_2 + G, \tag{14.2.5}$$

and from the boundary conditions (14.2.2) we have

$$B = \varphi^0/h, \quad G = 0. \tag{14.2.6}$$

Since the electrode plates are conductors, $D_2 = 0$ in the plates. Substituting from the solution (14.2.5) and (14.2.6) into (14.2.3), we obtain

$$D_2 = -\varepsilon\frac{\varphi^0}{h}. \tag{14.2.7}$$

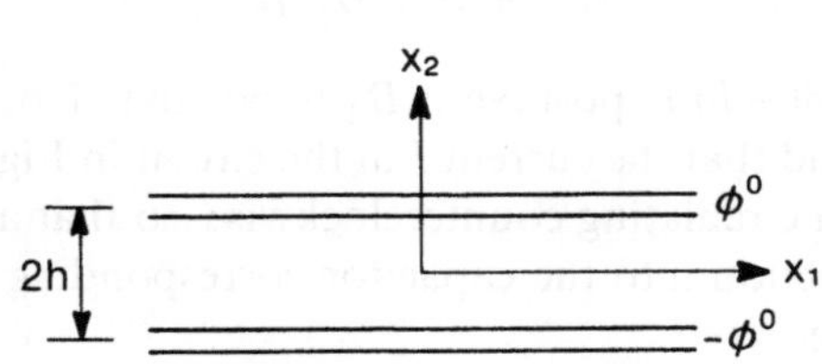

FIGURE 38. Parallel plate capacitor.

We also have the jump condition in (3.5.11), which is of the form

$$\mathbf{n}\cdot(\mathbf{D}^+ - \mathbf{D}^-) = 4\pi\sigma, \tag{14.2.8}$$

where σ is the surface charge in Gaussian units. Since at the top surface $n_i = \delta_{i2}$ and $\mathbf{D}^+ = 0$, from (14.2.7) and (14.2.8) we have

$$\sigma(h) = +\frac{\varepsilon\varphi^0}{4\pi h}, \tag{14.2.9}$$

and at the bottom surface $n_i = -\delta_{i2}$ and $\mathbf{D}^+ = 0$, from (14.2.7) and (14.2.8) we have

$$\sigma(-h) = -\frac{\varepsilon\varphi^0}{4\pi h}. \tag{14.2.10}$$

Since the voltage V across the capacitor is given by

$$V = \varphi^T - \varphi^B = \varphi^0 - (-\varphi^0) = 2\varphi^0, \tag{14.2.11}$$

from (14.2.9)–(14.2.11) we have

$$V = \pm\frac{4\pi 2h}{\varepsilon}\sigma(\pm h). \tag{14.2.12}$$

The total charge Q on the upper and lower surfaces of the electrode plates is given by

$$Q(\pm h) = \sigma(\pm h)A, \tag{14.2.13}$$

which with (14.2.12) enables us to write

$$V = \pm\frac{4\pi 2h}{\varepsilon A}Q(\pm h). \tag{14.2.14}$$

The quantity

$$\frac{\varepsilon A}{4\pi 2h} = C, \tag{14.2.15}$$

is called the capacitance of the capacitor. It is written in Gaussian units. From (14.2.7), (14.2.9) and (14.2.10) it may be noted that

$$\dot{\sigma}(\pm h) = \mp\dot{D}_2/4\pi, \tag{14.2.16}$$

which means that $\dot{\sigma}(-h)$ is positive if $\dot{D}_2$ is positive. Note further that the x_2 axis points up and that the current I in the circuit in Figure 37 is assumed to be positive when circulating counterclockwise, so that a positive I travels up in the $+x_2$-direction into the capacitor corresponding to a positive (up) $\dot{D}_2$. Hence, we have

$$I = \dot{\sigma}(-h)A = \dot{Q}(-h). \tag{14.2.17}$$

We now note that in the counterclockwise integral of $\mathbf{E}\cdot d\mathbf{r}$ that we will be performing for the circuit in Figure 37, we have

$$\int_B^T E\cdot dr = -\int_B^T \varphi_{,2}\, dX_2 = -[\varphi]_B^T = -\varphi^0 - \varphi^0 = -V, \quad (14.2.18)$$

and from (14.2.14), (14.2.15) and (14.2.17) we see that the relation for the capacitor that is needed in the circuit is

$$V = -Q(-h)/C. \quad (14.2.19)$$

Since the wire in the circuit in Figure 37 carries a current $\mathbf{J}$, we have to consider the counterclockwise line integral of $\mathbf{E}\cdot d\mathbf{r}$ due to the resistance in the wire. From $(10.2.7)_3$, we have

$$\mathbf{E} = \mathbf{J}/\sigma, \quad (14.2.20)$$

where σ is the electrical conductivity of the wire. Thus, for the wire carrying the current, we write

$$\int \mathbf{E}\cdot d\mathbf{r} = \int \frac{1}{\sigma}\mathbf{J}\cdot d\mathbf{r}, \quad (14.2.21)$$

where, for constant current density $\mathbf{J}$, we have

$$\mathbf{J} = \mathbf{n}I/a, \quad (14.2.22)$$

in which a is the area of the wire and $\mathbf{n}$ denotes a unit vector directed along the wire. Hence, from (14.2.21) and (14.2.22) we have

$$\int \mathbf{E}\cdot d\mathbf{r} = \int \frac{1}{\sigma}\frac{I}{a}\mathbf{n}\cdot d\mathbf{r}, \quad (14.2.23)$$

which, for constant a and σ, may be written

$$\int \mathbf{E}\cdot d\mathbf{r} = Il/\sigma a, \quad (14.2.24)$$

where l represents the length of the resistive region. Thus, we may write

$$\int \mathbf{E}\cdot d\mathbf{r} = IR, \quad (14.2.25)$$

where

$$R = l/\sigma a, \quad (14.2.26)$$

is the Ohmic resistance of the linear circuit.

14.3 Mutual and Self-Inductance

As noted at the beginning of Sec. 14.2 the equations for all current carrying portions of the circuits are Eqs. (10.1.7) and (10.1.21), which for convenience we write here in the form

$$c\nabla \times \mathbf{H} = 4\pi \mathbf{J}, \tag{14.3.1}$$

$$c\oint_C \mathbf{E}\cdot d\mathbf{r} = -\frac{\partial}{\partial t}\int_S \mathbf{n}\cdot\mathbf{B}\,dS. \tag{14.3.2}$$

We also have the linear constitutive equation $(10.2.7)_2$, i.e.,

$$\mathbf{B} = \mu\mathbf{H}, \tag{14.3.3}$$

which with (13.1.15) enables (14.3.1) to be written in the form

$$\nabla \times \nabla \times \mathbf{A} = \frac{4\pi\mu}{c}\mathbf{J}. \tag{14.3.4}$$

If we set $\nabla\cdot\mathbf{A} = 0$, Eq. (7.3.8) enables us to write (14.3.4) in the form

$$\nabla^2\mathbf{A} = -\frac{4\pi\mu}{c}\mathbf{J}, \tag{14.3.5}$$

from which on account of (6.4.6) and (6.3.12), we obtain

$$\mathbf{A} = \frac{4\pi\mu}{c}\int\frac{\mathbf{J}\,dV}{r_1}. \tag{14.3.6}$$

The result in (14.3.6) holds for any number of circuits where the integral simply runs over the volume occupied by all the circuits. If all the circuits are composed of the same material, μ is a constant. If it varies from circuit to circuit, its variation must be included. Thus for N circuits we write

$$\mathbf{A}^{(p)} = \frac{4\pi}{c}\sum_n^N \mu^{(n)}\int_{V^{(n)}}\frac{\mathbf{J}^{(n)}\,dV^{(n)}}{r_1^{(n)(p)}}, \tag{14.3.7}$$

for $\mathbf{A}$ at the pth circuit. For all N circuits except the pth, we may make the replacement

$$\mathbf{J}^{(n)}\,dV^{(n)} = I^{(n)}\,d\mathbf{s}^{(n)}, \tag{14.3.8}$$

which treats all current in the circuit as if it is concentrated at the centerline. The current $I^{(n)}$ does not vary with position in the nth circuit, since we are considering only series circuits here, but does vary with time. It should be noted that in making the replacement in (14.3.8) we have permitted $\mathbf{J}$ to become unbounded, and we require boundedness for $\mathbf{J}$ in any region in which $\mathbf{A}$ is defined. This is the reason that the replacement in (14.3.8) is not permissible in the pth circuit in which $\mathbf{A}$ in (14.3.7) is defined. For uniform current in the pth circuit of cross-sectional area $a^{(P)}$, we write

$$\mathbf{J}^{(p)}\,dV^{(p)} = \frac{I^{(p)}}{a^{(p)}}\,da^{(p)}\,d\mathbf{s}^{(p)}, \tag{14.3.9}$$

in place of (14.3.8), which with (14.3.8) enables us to write (14.3.7) in the form

$$\mathbf{A}^{(p)} = \frac{4\pi}{c} \sum_{n \neq p}^{N} \mu^{(n)} I^{(n)} \oint_{C^{(n)}} \frac{d\mathbf{s}^{(n)}}{r_1^{np}} + \frac{4\pi}{c} \mu^{(p)} \frac{I^{(p)}}{a^{(p)}} \oint_{C^{(p)}} \int_{a^{(p)}} \frac{d\mathbf{s}^{(p)}\, da^{(p)}}{r_1^{pp}}, \qquad (14.3.10)$$

in which all the integrals are convergent.

We have not as yet employed Faraday's law (14.3.2). We now substitute from (13.1.16) into (14.3.2), which we write for the pth circuit in the form

$$c \oint_{C^{(p)}} \mathbf{E}^{(p)} \cdot d\mathbf{r} = -\frac{\partial}{\partial t} \int_{S^{(p)}} \mathbf{n} \cdot \nabla \times \mathbf{A}^{(p)}\, dS, \qquad (14.3.11)$$

where $S^{(p)}$ is *an* open area enclosed by the centerline of the pth circuit. With the aid of Stokes theorem (14.3.11) yields

$$c \oint_{C^{(p)}} \mathbf{E}^{(p)} \cdot d\mathbf{r}^{(p)} = -\frac{\partial}{\partial t} \oint_{C^{(p)}} \mathbf{A}^{(p)} \cdot d\mathbf{r}^{(p)}, \qquad (14.3.12)$$

and substituting for $\mathbf{A}^{(p)}$ from (14.3.10) we obtain

$$\oint_{C^{(p)}} \mathbf{E}^{(p)} \cdot d\mathbf{r}^{(p)} = -\frac{4\pi}{c^2} \sum_{n \neq p}^{N} \mu^{(n)} \dot{I}^{(n)} \oint_{C^{(p)}} \oint_{C^{(n)}} \frac{d\mathbf{r}^{(p)} \cdot d\mathbf{s}^{(n)}}{r_1^{np}}$$
$$-\frac{4\pi \mu^{(p)}}{c^2 a^{(p)}} \dot{I}^{(p)} \oint_{C^{(p)}} \oint_{C^{(p)}} \int_{a^{(p)}} \frac{d\mathbf{r}^{(p)} \cdot d\mathbf{s}^{(p)}}{r_1^{pp}}\, da^{(p)}, \qquad (14.3.13)$$

since the $I^{(n)}$ are the only quantities that vary in time because the geometry and magnetic susceptibilities are fixed. If, as is usually the case, the same material is used for each circuit $\mu^{(n)} = \mu$ and may be taken out of the summation sign. Clearly, we may write

$$L_{pn} = \frac{4\pi}{c^2} \mu^{(n)} \oint_{C^{(p)}} \oint_{C^{(n)}} \frac{d\mathbf{r}^{(p)} \cdot d\mathbf{s}^{(n)}}{r_1^{np}}, \qquad (14.3.14)$$

where the L_{pn} are called the coefficients of mutual inductance for $n \neq p$, and may be calculated by evaluating the integral in (14.3.14) since we have assumed that two circuits never touch. Similarly, we may write

$$L_{pp} = \frac{4\pi \mu^{(p)}}{c^2 a^{(p)}} \oint_{C^{(p)}} \oint_{C^{(p)}} \frac{d\mathbf{r}^{(p)} \cdot d\mathbf{s}^{(p)}}{r_1^{pp}}\, da^{(p)}, \qquad (14.3.15)$$

where the L_{pp} are called the coefficients of self-inductance, and may be calculated by evaluating the integral in (14.3.15). With the definitions in (14.3.14) and (14.3.15), Eq. (14.3.13) may be written in the form

$$\oint_{C^{(p)}} \mathbf{E}^{(p)} \cdot d\mathbf{r}^{(p)} = -\sum_{n}^{N} L_{pn} \dot{I}^{(n)}, \qquad (14.3.16)$$

in which n may equal p, but L_{pp} is defined in (14.3.15) while the other L_{pn} are defined in (14.3.14).

We have evaluated the integral on the left-hand side of (14.3.16) across a generator in Sec. 14.1 and across a capacitor and a resistor in Sec. 14.2, and we have expressed the integral in terms of circuit parameters in (14.1.6), (14.2.19) with (14.2.18) and (14.2.25) for a generator, capacitor and resistor, respectively. Collecting all these integrals, we may write

$$\oint_C \mathbf{E}^{(p)} \cdot d\mathbf{r} = -\mathscr{E} + I^{(p)}R^{(p)} + \frac{Q^{(p)}(-h)}{C^{(p)}}. \qquad (14.3.17)$$

Thus, substituting from (14.3.17) into (14.3.16), we finally obtain the circuit equation for the pth circuit, in the form

$$-\mathscr{E}^{(p)} + I^{(p)}R^{(p)} + \frac{Q^{(p)}(-h)}{C^{(p)}} = -\sum_n L_{pn}\dot{I}^{(n)}, \qquad (14.3.18)$$

which with (14.2.17) may be written

$$\sum_n L_{pn}\ddot{Q}^{(n)} + R^p\dot{Q}^{(p)} + \frac{Q^p}{C^{(p)}} = \mathscr{E}^{(p)}, \qquad (14.3.19)$$

as we intended to show.

14.4 Parallel Circuits

It should be noted that it is perfectly clear from the development in Sec. 14.3 that for circuits in parallel as shown in Figure 39 each current carrying element is counted *once* in the integral for $\mathbf{A}^{(p)}$ in (14.3.10) for the determination of the self- and mutual-inductances. The only distinction is that different currents flow in different arms of the loop integrals in (14.3.10)–(14.3.13) in the case of parallel circuits. With this modification, the line integrals in (14.3.12) around each *independent* closed loop determine all the Kirchhoff voltage balance equations for the system.

The Kirchhoff current balance equations at each junction are determined from the equation of the conservation of charge when there is no net charge, i.e., (6.3.23), which may also be obtained by taking the divergence of (14.3.1) which yields

$$\nabla \cdot \mathbf{J} = 0. \qquad (14.4.1)$$

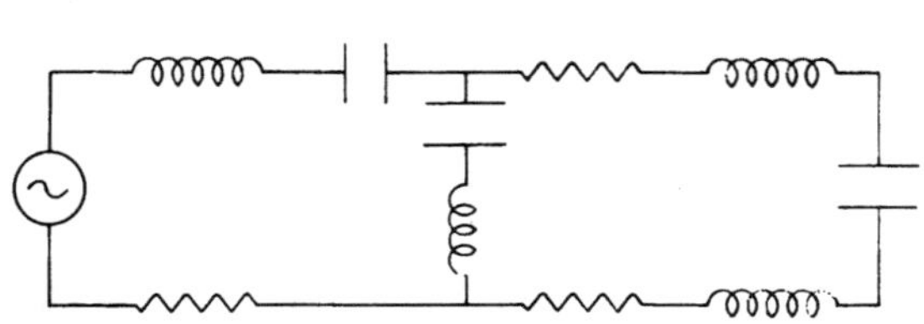

FIGURE 39. Parallel circuit.

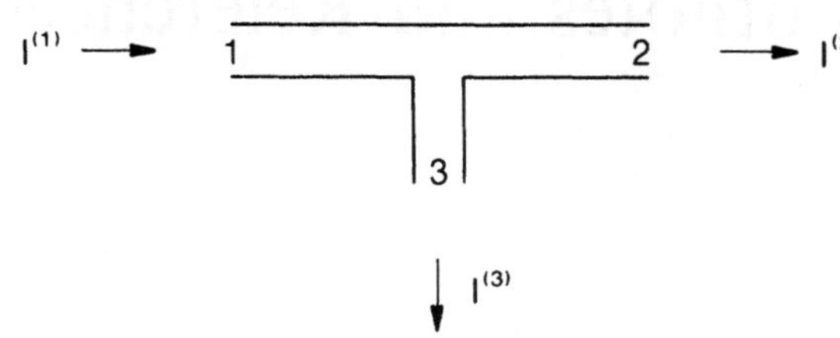

FIGURE 40. Enlarged diagram of circuit junction.

For a finite current carrying region application of the divergence theorem yields

$$\int_V \nabla\cdot\mathbf{J}\,dV = \int_S \mathbf{n}\cdot\mathbf{J}\,dS = 0. \tag{14.4.2}$$

Applying (14.4.2), e.g., to the junction shown in Figure 40 we obtain

$$\int_{A_1} \mathbf{n}^{(1)}\cdot\mathbf{J}^{(1)}\,dS + \int_{A_2} \mathbf{n}^{(2)}\cdot\mathbf{J}^{(2)}\,dS + \int_{A_3} \mathbf{n}^{(3)}\cdot\mathbf{J}^{(3)}\,dS = 0, \tag{14.4.3}$$

since the current densities are confined to the wires. Since for the positively assumed $I^{(n)}$ in the diagram, we have

$$\mathbf{J}^{(1)} = -\mathbf{n}^{(1)}J^{(1)}, \quad \mathbf{J}^{(2)} = \mathbf{n}^{(2)}J^{(2)}, \quad \mathbf{J}^{(3)} = \mathbf{n}^{(3)}J^{(3)}, \tag{14.4.4}$$

where the $\mathbf{n}^{(m)}$ denote the outwardly directed unit normals, we obtain

$$-I^{(1)} + I^{(2)} + I^{(3)} = 0, \tag{14.4.5}$$

as the Kirchhoff current balance equation at the junction. Clearly, a similar analysis holds for any number of wires connected at a junction.

Footnotes and References

1. In microscopic physics the particles are modeled as having *all* matter located at one mathematical point, which must be the centroid if the particle actually has any dimension at all, no matter how small. The point particle is a convenient model, but it is only a model.
2. Although I cannot locate this precise quotation, which I am certain I have read in this form, a lengthier discussion containing the same information may be found on pages 35 and 37 of *Albert Einstein: Philosopher–Scientist*, Vol. I, edited by Paul Arthur Schilpp (Harper and Brothers, 1959).
3. G.H. Livens, *The Theory of Electricity* (Cambridge University Press, London, 1926), second ed.
4. Max Mason and Warren Weaver, *The Electromagnetic Field* (Dover Publications, Inc., New York, by permission of the University of Chicago Press, 1929).
5. H.F. Tiersten and C.F. Tsai, "On the Interaction of the Electromagnetic Field with Heat Conducting Deformable Insulators," J. Math. Phys. **13**, 361 (1972).
6. C. Truesdell and R.A. Toupin, "The Classical Field Theories," in *Encyclopedia of Physics*, edited by S. Flügge (Springer-Verlag, Berlin, 1960).
7. A. Einstein, "On the Electrodynamics of Moving Bodies," translated from "Zur Elektrodynamik bewegter Körper," Ann. Phys. **17** (1905) in *The Principle of Relativity*, translated by W. Perrett and G.B. Jeffrey (Dover, New York, by permission of Methuen and Company, Ltd., 1923), pp. 35–65.
8. Similar considerations are performed in time in order to eliminate time fluctuations and obtain smooth fields in time. This is assumed to have been done and is not of direct interest here.
9. H.B. Phillips, *Vector Analysis* (John Wiley and Sons, Inc., New York, 1950), pp. 114–115.
10. We have shown that Eqs. (2.5.6) and (2.5.7) hold if ρ and its first spatial derivatives are continuous. It can be shown that they hold if ρ is continuous and its first spatial derivatives are piecewise continuous. See Ref. 9, Sec. 59.
11. The Section is strongly influenced by Ref. 3, Secs. 61–63.
12. This is a result of the fact that the r_1^2 in the numerator on account of (2.3.2) cancels the r_1^2 in the denominator resulting from (3.2.2) in the integral over Σ.
13. At this stage in the development of an electrostatic field theory inside a region containing a continuous distribution of quadrople density, a quantity much larger (infinite) than that retained would have to be ignored. It seems to me that a theory resulting from such an approach would be meaningless.

14. In fact if s separates an insulator from a conductor ρ must diverge at s.
15. Reference 9, Sec. 64.
16. In fact, in the same way that the jump in the normal component of **D** across a metal-insulator interface, such as exists in a capacitor, gives the charge on the capacitor plates, the jump in the normal component of $\mathbf{T}^E$ across the interface gives the mechanical force between the capacitor plates.
17. Reference 9, Eq. (458). The integral theorem used in going from (2.4.13) to (2.4.14) is of the same nature and is given in Eq. (124) of Ref. 9.
18. Reference 9, Eq. (122).
19. Reference 9, Eq. (125).
20. Reference 9, Eq. (459).
21. Reference 9, Eq. (123).
22. H. Jeffreys, *Cartesian Tensors* (Cambridge University Press, 1952), Chap. I.
23. Reference 9, Eq. (456).
24. In this discretization, as in all the discretizations of loop integrals in magnetostatics, the number of current elements $Y^{(n)}$ is countably infinite. This is a consequence of the fact that the line loop constitutes a linear continuum and offers no conceptual or mathematical difficulty in the model.
25. The general case is treated in the next section.
26. Although the result in (9.1.19) has been derived under a restriction on the current density, the restriction is removed in the next section where it is shown that the result holds in general.
27. The expression in Eq. (10.2.5) actually represents the rate of entropy production and one cannot say whether it increases the temperature locally or causes heat conduction or what combination thereof without the appropriate thermodynamic considerations.
28. A.A. Michelson and E.W. Morley, "Influence of Motion of the Medium on the Velocity of Light," Amer. J. Sci. **31**, 377 (1886).
29. This requires a simple proper orthogonal transformation from this special orthogonal coordinate system to an arbitrary orthogonal coordinate system. See C. Möller, *The Theory of Relativity* (Oxford University Press, 1952), Sec. 18.
30. See C. Möller, *The Theory of Relativity* (Oxford University Press, 1952), Secs. 33, 52 and 53. Lorentz transformations turn out to be orthogonal transformations in the four-dimensional space-time continuum introduced by Minkowski.
31. There are many interesting effects resulting from the Lorentz transformation. See Möller or any text that discusses relativity.
32. This has been done before in one book that I know of, although in somewhat less detail. P. Bergmann, *Basic Theories of Physics, Mechanics and Electrodynamics* (Prentice-Hall, Inc., 1949), Sec. 9.3.
33. It is a simple matter to obtain the relativistic-transformation, relations that are inverse to (12.3.19) and (12.3.32)–(12.3.34), from which all others can be obtained, but it takes space and is not deemed to be worthwhile for our purposes.
34. In my opinion this approximation is more useful than the completely general relativistic procedure because of its purely formal nature. The approximate procedure is more useful because it is more flexible in that it readily permits the use of models in describing diverse material behavior. This type of approximation is employed routinely in solid state physics.

35. We mean not magnetizable to lowest order, which in this case means linearly magnetizable as a result of the (here) negligible magnetic induction field. It is not considered purposeful to include this nuance here.
36. It should be noted that we are considering any gauge in which φ does not vanish.
37. A comment similar to 35, but with polarizable replacing magnetizable and electric field replacing magnetic induction field.

Bibliography

Electromagnetism

J.C. Maxwell, *A Treatise on Electricity and Magnetism*, Vol. I and II (Dover Publications, Inc., New York, 1954).

W.K.H. Panofsky and M. Phillips, *Classical Electricity and Magnetism* (Addison-Wesley Publishing Company, Inc., Reading, Massachusetts, 1955).

R. Becker and F. Sauter, *Electromagnetic Fields and Interactions*, Vol. I (Blaisdell, New York, 1964).

D.M. Cook, *The Theory of the Electromagnetic Field* (Prentice-Hall, Inc., Englewood Cliffs, New Jersey, 1975).

Interactions with Deformable Continua

W.F. Brown, Jr., *Magnetoelastic Interactions* (Springer-Verlag, Berlin, 1966).

R.N. Thurston, "Waves in Solids," in *Encyclopedia of Physics*, edited by C. Truesdell (Springer-Verlag, Berlin, 1974), Vol. VIa/4.

Y.H. Pao, "Electromagnetic Forces in Deformable Continua," in *Mechanics Today*, edited by S. Nemat-Nasser (Pergamon Press, Oxford, 1978), Vol. 4, pp. 209–305.

K. Hutter and A.A.F. van de Venn, *Field Matter Interactions in Thermoelastic Solids* (Springer-Verlag, Berlin, 1978).

D.F. Nelson, *Electric, Optic and Acoustic Interactions in Dielectrics* (John Wiley and Sons, New York, 1979).

F.C. Moon, *Magneto-Solid Mechanics* (John Wiley and Sons, New York, 1984).

G.A. Maugin, *Continuum Mechanics of Electromagnetic Solids* (North-Holland, New York, 1988).

Author Index

Subject Index

Springer Tracts in Natural Philosophy

Vol. 1 Gundersen: Linearized Analysis of One-Dimensional Magnetohydrodynamic Flows. With 10 figures. X, 119 pages. 1964.

Vol. 2 Walter: Differential- und Integral-Ungleichungen und ihre Anwendung bei Abschätzungs- und Eindeutigkeitsproblemen
Mit 18 Abbildungen. XIV, 269 Seiten, 1964.

Vol. 3 Gaier: Konstruktive Methoden der konformen Abbildung
Mit 20 Abbildungen und 28 Tabellen. XIV, 294 Seiten. 1964.

Vol. 4 Meinardus; Approximation von Funktionen und ihre numerische Behandlung
Mit 21 Abbildungen. VIII, 180 Seiten, 1964.

Vol. 5 Coleman, Markovitz, Noll: Viscometric Flows of Non-Newtonian Fluids. Theory and Experiment
With 37 figures. XII, 130 pages. 1966.

Vol. 6 Eckhaus: Studies in Non-Linear Stability Theory
With 12 figures. VIII, 117 pages. 1965.

Vol. 7 Leimanis: The General Problem of the Motion of Coupled Rigid Bodies About a Fixed Point
With 66 figures, XVI, 337 pages. 1965.

Vol. 8 Roseau: Vibrations non linéaires et théorie de la stabilité
Avec 7 figures. XII, 254 pages. 1966.

Vol. 9 Brown: Magnetoelastic Interactions
With 13 figures. VIII, 155 pages. 1966.

Vol. 10 Bunge: Foundations of Physics
With 5 figures. XII, 312 pages. 1967.

Vol. 11 Lavrentiev: Some Improperly Posed Problems of Mathematical Physics
With 1 figure. VIII, 72 pages. 1967.

Vol. 12 Kronmüller: Nachwirkung in Ferromagnetika
Mit 92 Abbildungen. XIV, 329 Seiten. 1968.

Vol. 13 Meinardus: Approximation of Functions: Theory and Numerical Methods
With 21 figures. VIII, 198 pages. 1967.

Vol. 14 Bell: The Physics of Large Deformation of Crystalline Solids
With 166 figures. X, 253 pages. 1968.

Vol. 15 Buchholz: The Confluent Hypergeometric Function with Special Emphasis on its Applications
XVIII, 238 pages. 1969.

Vol. 16 Slepian: Mathematical Foundations of Network Analysis
XI, 195 pages. 1968.

Vol. 17 Gavalas: Nonlinear Differential Equations of Chemically Reacting Systems
With 10 figures. IX, 107 pages. 1968.

Vol. 18 Marti: Introduction to the Theory of Bases
XII, 149 pages. 1969.

Vol. 20 Edelen/Wilson: Relativity and the Question of Discretiztion in Astronomy
With 34 figures. XII, 186 pages. 1970.

Vol. 21 McBride: Obtaining Generating Functions
XIII, 100 pages. 1971.

Vol. 22 Day: The Thermodynamics of Simple Materials with Fading Memory
With 8 figures. X, 134 pages. 1972.

Vol. 23 Stetter: Analysis of Discretization Methods for Ordinary Differential Equations
With 12 figures. XVI, 388 pages. 1973.

Vol. 24 Strieder/Aris: Variational Methods Applied to Problems of Diffusion and Reaction
With 12 figures, IX, 109 pages. 1973.

Vol. 25 Bohl: Momotonie: Losbarkeit und Numerik bei Operatorgleichungen
Mit 9 Abbildungen. IX, 255 Seiten. 1974.

Vol. 26 Romanov: Integral Geometry and Inverse Problems for Hyperbolic Equations
With 21 figures. VI, 152 pages. 1974.

Vol. 27 Joseph: Stability of Fluid Motions I
With 57 figures. XIII, 282 pages. 1976.

Vol. 28 Joseph: Stability of Fluid Motions II
With 39 figures. XIV, 274 pages. 1976.

Vol. 29 Bressan: Relativistic Theories of Materials
XIV, 290 pages. 1978.

Vol. 30 Day: Heat Conduction within Linear Thermoelasticity
VII, 82 pages. 1985.

Vol. 31 Valent: Boundary Value Problems of Finite Elasticity
XIII, 191 pages. 1988.

Vol. 32 Day: A Commentary on Thermodynamics
IX, 96 pages. 1988.

Vol. 33 Cohen/Muncaster: The Theory of Pseudo-rigid Bodies
X, 180 pages. 1988.

Vol. 34 Angeles: Rational Kinematics
XII, 137 pages. 1989.

Vol. 35 Capriz: Continua with Microstructure
X, 92 pages. 1989.

Vol. 36 Tiersten: A Development of the Equations of Electromagnetism
in Material Continua
With 40 figures. XI, 156 pages. 1990.

MIX
Papier aus verantwortungsvollen Quellen
Paper from responsible sources
FSC® C105338

If you have any concerns about our products,
you can contact us on
ProductSafety@springernature.com

In case Publisher is established outside the EU,
the EU authorized representative is:
Springer Nature Customer Service Center GmbH
Europaplatz 3, 69115 Heidelberg, Germany

Printed by Libri Plureos GmbH
in Hamburg, Germany